Food Quality Handbook

Food Quality Handbook

Edited by **Margo Field**

New York

Published by Callisto Reference,
106 Park Avenue, Suite 200,
New York, NY 10016, USA
www.callistoreference.com

Food Quality Handbook
Edited by Margo Field

International Standard Book Number: 978-1-63239-341-8 (Hardback)

Printed in the United States of America.

Contents

Preface

The purpose of the book is to provide a glimpse into the dynamics and to present opinions and studies of some of the scientists engaged in the development of new ideas in the field from very different standpoints. This book will prove useful to students and researchers owing to its high content quality.

The quality characteristics of food that are satisfactory to the consumers are collectively known as food quality. This text describes important scientific avenues for the development and growth of food quality and also offers food scientists with relevant information for future advancements. Food analysts, scientists and even research workers will benefit from the elaborated methods and experimental information provided in the book and industrial experts can use the information as a noteworthy reference source. The book can most aptly be used as an instrument by analysts for increasing their knowledge with the latest scientific data for quality assessment. Case studies in the book present knowledge on the development of food quality in marine and land creatures in natural habitat.

At the end, I would like to appreciate all the efforts made by the authors in completing their chapters professionally. I express my deepest gratitude to all of them for contributing to this book by sharing their valuable works. A special thanks to my family and friends for their constant support in this journey.

Editor

Section 1

Molecular Approaches to Achieve the Food Quality

1

Monitoring Harmful Microalgae by Using a Molecular Biological Technique

Tomotaka Shiraishi[1], Ryoma Kamikawa[2], Yoshihiko Sako[3] and Ichiro Imai[4]
[1]*Wakayama Research Center of Agriculture, Forestry and Fisheries,*
[2]*University of Tsukuba,*
[3]*Kyoto University,*
[4]*Hokkaido University,*
Japan

1. Introduction

In recent years, cultivation of fish and shellfish possesses an important portion for securing enough seafood all over the world. While, fisheries industry handling fish and shellfish derived from cultivation in addition to natural seafood are exposed to the danger of mass mortality of the reared and toxicities of bivalves, sometimes resulting in serious economic losses and physiological damages by seafood poisoning.

Certain microalgal species have been clearly demonstrated relationships with a mass mortality of fish and shellfish and certain symptoms of people which are caused by consumption of seafood contaminated with toxins. Occurrences of paralytic shellfish poisoning (PSP), neurotoxic shellfish poisoning (NSP), diarrheic shellfish poisoning (DSP), amnesic shellfish poisoning (ASP) and ciguatera fish poisoning (CFP) are caused through a food chain from toxin-producing microalgae to fish or shellfish (Hallegraeff 1995). Otherwise, some microalgal species cause a red tide, the name commonly used for the occurrence of harmful algal blooms (HABs) that result from local or regional accumulation of a unicellular phytoplankton species and exert a negative effect on the environment (Anderson 1994; Smayda 1997). Of the 5000 species of extant marine phytoplankton, approximately 300 algal species can form red tides, and the distribution of these HAB species is increasing globally. HABs therefore continue to receive attention in coastal regions all over the world (Hallegraeff 1993).

The canonical method monitoring HABs is that by observation of morphological features under a light microscope. This method requires labour, time, expert knowledge on morphologies of microalgae, and technical skills to observe the species-specific morphological features. In addition, morphology of microalgae is sometimes changed, depending on the environmental conditions or their growth phases (Imai 2000). Therefore, identification of HAB species with ambiguous morphology is quite difficult and sometimes subjective, and henceforth, problematic particularly in genera comprising both toxic and non-toxic species which have similar morphology. The difficulty of monitoring HAB by light microscopy has indicated necessity of a more objective, rapid and accurate identification method for HAB species.

In the last decade, to address the above issue, molecular biological techniques have been developed for monitoring HAB species (Godhe et al. 2002; Sako et al. 2004; Hosoi-Tanabe and Sako 2005a). Many of such newly developed methods focus on genetic diversity of a certain gene that does not change in a short term depending on environmental conditions or the algal growth phase. This implies such molecular biological techniques can distinguish a HAB species from a morphologically similar but non-toxic species if the two species have gene sequences different from each other. Additionally, these assays appear to be time-saving, accurate, simple, and effective for the mass investigation of samples. So, polymerase chain reaction (PCR) assay, one of representative molecular biological techniques, is an indispensable tool in the fields regarding HAB-monitoring, since a PCR-based method allows us to identify or detect HAB cells more objectively even if they have morphology difficult to be defined their taxonomy under general microscopic methods (Adachi et al. 1994). Further, real-time PCR assay was later developed which allows us not only to detect and identify HAB species but also to quantify HAB cells (Bowers et al. 2000).

This is especially useful for resting cysts of certain HAB species. Some HAB species have a resting stage like a seed as one of their life cycles, and resting cysts of many HAB species are spheroid or ovate and with neither species-specific colour nor ornament (e.g., *Alexandrium*;).

In the chapter, we will introduce the principle of real-time PCR itself at first, and subsequently focus on several applications of the real-time PCR assay which have been developed to monitor dynamics of HAB species: e.g., neurotoxin-producing dinoflagellate *Alexandrium* species, red tide-forming dinoflagellates *Karenia mikimotoi* and *Cochlodinium polykrikoides*, red tide-forming raphidophytes *Chattonella* species and *Heterosigma akashiwo*, and bivalve-specific killer dinoflagellate *Heterocapsa circularisquama*.

Further, we introduce a method I and my coworkers have recently developed to process a lot of environmental seawater samples by using a filtration assay and the simplest protocol of DNA extraction (Shiraishi et al. 2009.). The simple method allows us to investigate many seawater samples for monitoring HAB species smoothly by using a real-time PCR assay with HAB species specific oligonucleotide primers and a probe.

2. Principle and application of real-time PCR assay for harmful algal blooms

In 1991, Holland et al. (1991) developed the new method called "*Taq* man real-time PCR assay", which is based on the 5′ to 3′-exonuclease activity of *Taq* polymerases, for monitoring the quantity of PCR product in real-time. Subsequently, the method was improved by Heid et al. (1996). The feature of the real-time PCR assay is requirement of a fluorogenic oligonucleotide probe in addition to reagents used in general PCR-based assay. The emission of 6-carboxy-tetramethyl-rhodamine (TAMRA) attached at the 3′-termini of probes as a quencher dye suppresses that of 6-carboxy-fluorescein (FAM) attached at the 5′-termini as a reporter dye due to the proximity between the emissions of two dyes. Describing the mechanisms of quantification briefly, the labeled probes hybridize with target DNA or PCR products and subsequently are deleted by the exonuclease activity in each PCR-cycle, resulting in release of emission of the reporter dye. A fluorometer, which is generally equipped with a thermal cycler, detects the released emission of the reporter dye and quantifies the PCR products. Due to the utility, high-sensitivity, and accuracy of quantification, real-time PCR assay has been applied to development of a method for monitoring several HABs.

The first application of *Taq* man real-time PCR assay to HAB species was performed by Bowers et al. (2000) to quantitatively detect the toxic dinoflagellate *Pfiesteria piscicida* and its close relative *Pfiesteria shumwayae*. Bowers et al. (2000) designed primers-*Taq* man probe sets specifically hybridizes 18S rRNA gene of either *P. piscicida* or *P. shumwayae*. The real-time PCR assay using the primers-probe sets demonstrated high specificity even for single cells. Similar trials were carried out for the toxic dinoflagellate *Alexandrium* species (Galluzzi et al., 2004; Hosoi-Tanabe and Sako, 2005; Dyhrman et al. 2006), *Karenia brevis* (Gray et al. 2003), *Pfiesteria* spp. (Zhang and Lin, 2005), the naked harmful dinoflagellate *Cochlodinium polykrikoides, Karenia mikimotoi* (Kamikawa et al., 2006), and harmful raphidophytes (Handy et al., 2005; Bowers et al. 2006; Kamikawa et al., 2006).

Especially, *Taq* man real-time PCR assay was applied to resting cysts of *Alexandrium* species in marine sediments (Kamikawa et al. 2005, 2007; Erdner et al. 2011). The cyst densities calculated by the real-time PCR assay for *Alexandrium* cysts were almost identical to those by the canonical method to monitoring the cysts called primulin-staining (Yamaguchi et al. 1995, ; Kamikawa et al. 2007). However, it is notable that the cyst density calculated by the real-time PCR assay tends to be lower than that by the primulin method when sediment samples collected from 1-3cm depth were used (Erdner et al. 2011). This difference between the real-time PCR assay and the primulin method suggests that the real-time PCR assay may be influenced by cyst condition and viability (Erdner et al. 2011). Otherwise, there are unknown species that produce resting cysts with the similar morphology and that are stained with primulin as well.

3. The noxious dinoflagellate *Heterocapsa circularisquama*

The dinoflagellate *Heterocapsa circularisquama* is one of the most noxious phytoplankton in Japanese coastal areas and causes mass mortalities of both natural and cultured bivalves such as oyster, manila clam and pearl oyster in Japan (Nagai et al., 1996, 2000; Matsuyama, 1999). Blooms of this species have had significant negative impacts on the shellfish aquaculture especially in western coastal area of Japan (Matsuyama et al., 1997; Tamai, 1999). *H. circularisquama* was discovered for the first time in Uranouchi Inlet, Kochi Prefecture, Japan in 1988, and since that time, bloom occurrences have expanded throughout the western area of Japan (Matsuyama et al., 2001; Imai et al., 2006).

Monitoring the population dynamics of this species is essential for forecast of the red tide occurrences, and hence, for the mitigation of the damages, following to early countermeasures. Generalized seasonal occurrence of this species in summer and autumn could be determined using conventional optical microscopy (Matsuyama et al., 1996; Nakanishi et al., 1999; Shiraishi et al., 2007). However, precise identification and enumeration are difficult because this species is rather smaller than other red tide species (<30 μm), and there are numerous co-occurring dinoflagellates with similar morphology, implying that it is difficult to distinguish them from *H. circularisquama* (Horiguchi, 1995;

Iwataki et al., 2004; see also Fig. 1). Moreover, definitive identification of this species is based on morphology of body scales that can only be visualized using transmission electron microscopy (Horiguchi, 1995).

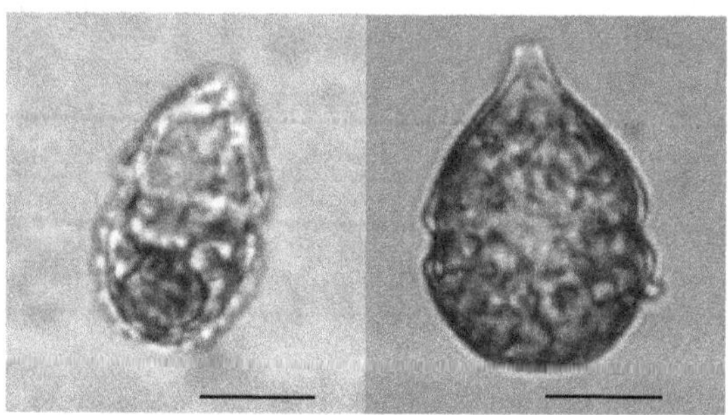

Fig. 1. Morphology of *Heterocapsa circularisquama* (left) and *Scrippsiella* sp. (right). Bar: 10μm.

Shiraishi et al. (2007) reported that it was possible to monitor *H. circularisquama* using an indirect fluorescent antibody technique (IFAT). This IFAT method allowed accurate detection of the cells even at low densities (lower limit, ca. 1 cell L^{-1}). Field studies using this method allowed the population dynamics of this species to be determined throughout a year in Uranouchi Inlet, Kochi Prefecture, Japan (Shiraishi et al. 2008.), and during early spring and later winter in Ago Bay, Mie Prefecture, Japan. Though the method demonstrated high specificity, individual *H. circularisquama* cells at lower density become difficult to be detected by epifluorescent microscopic observation in samples dominated by similar sized phytoplankton. This means additional treatments and significantly longer times were required to detect *H. circularisquama* cells in such samples. Consequently, there is still an urgent need to develop a simpler, quantitative method for monitoring *H. circularisquama*.

Kamikawa et al. (2006) previously reported a real-time PCR identification method of *H. circularisquama*. Though it could be used in the field, the assay as described requires a long and complex DNA extraction processes. Additionally, since target cells in cultures and seawater samples were collected by centrifugation, which imply that cells might be lost during the process and that only small sample volumes (50 mL at most) could be readily processed at a time. The conventional method using filtration was not feasible for concentrating *H. circularisquama* cells because most cells were attached and trapped on the surface of certain filters (Shiraishi et al., 2007). Thus, it was important to develop a simple technique for cell collection and DNA extraction to apply the *H. circularisquama*-specific real-time PCR assay to the field monitoring.

In our previous study, we developed a simple and quantitative monitoring method of *H. circularisquama* using a real-time PCR assay (Shiraishi et al. 2009.). The DNA extraction was performed within a relatively short time by gently filtering the cells down on a filter and then simply boiling the filter in a buffer. The population dynamics of *H. circularisquama* in an inlet revealed by the real-time PCR assay and by the IFAT assay were well consistent with each other. Because the method was only simply described in the original paper by the limitation of printing, we introduce the protocol of the simple real-time PCR assay (Shiraishi et al. 2009.) in detail in the following sections. This protocol will be helpful for the studies on many other dinoflagellate species by real-time PCR assay.

4. Materials and methods

4.1 Organisms and culture conditions

The algal strains of *H. circularisquama*, *K. mikimotoi* and *Skeletonema* sp. were obtained from the National Research Institute of Fisheries and Environment of Inland Sea, Fisheries Research Agency. Strains of *Heterocapsa triquetra* and *Heterosigma akashiwo* were isolated by G. Nishitani from Maizuru Bay, Kyoto Prefecture, Japan in 1998 and by I. Imai from Hiroshima Bay, Hiroshima Prefecture, Japan in 1989, respectively. These strains were cultured at a temperature of 20 °C on a 14-h light: 10-h dark photo-cycle under an illumination at 180 μmol photons m^{-2} s^{-1} in modified SWM-3 medium (Chen et al., 1969; Imai et al., 1996).

4.2 Cell collection and DNA extraction

The most effective method of extracting *H. circularisquama* DNA from cell pellets was evaluated using six different protocols, the simplest of which was a TE boiling method modified from the procedure of Kamikawa et al. (2006). For the basic boiling procedure, either one cell or 100 cells of cultured *H. circularisquama* cells were collected on the Nuclepore polycarbonate membrane filters (pore size 3.0 μm) (Whatman, Maidstone, UK) by filtration, respectively. Extractions at both cell concentrations were done in triplicate to assess assay variability. The filter was placed in a 1.5-mL microtube without folding, and then 750 μL of TE buffer (10 mM Tris-HCl: pH 8.0, 1 mM EDTA: pH 8.0) was added. After the boiling for 10 min in the TE buffer, the filter was immediately removed. The extracted DNA sample was stored at -60 °C until the real-time PCR assay was performed. DNA extraction efficiency using the TE (Tris-HCl/EDTA) boiling method was compared with that by the modified CTAB (Cetyltrimethylammonium Bromide) method (Zhou et al., 1999; Kamikawa et al., 2005) and the proteinase K method (Kamikawa et al., 2005), both of which are commonly used. The same protocol as described above for the TE extractions was followed.

To choose the most suitable filter for the DNA extraction of *H. circularisquama*, the cells were collected on 6 different filters and the DNA extraction was performed on each filter by the TE boiling method which was found to be the efficient method from the study described above. Specifically, one cell and 100 cells of cultured *H. circularisquama* cells were collected by filtration onto membrane filters composed of either polycarbonate membranes (Nuclepore, mesh size 3.0 μm), glass-fibers (GF/C, pore size 1.2 μm) (Whatman, Maidstone, UK), cellulose mixed esters (pore size 3.0 μm) (Millipore, Tokyo, Japan), cellulose acetate (pore size 3.0 μm) (ADVANTEC, Tokyo, Japan), polytetrafluoroethylene (PTFE) (pore size 3.0 μm) (ADVANTEC, Tokyo, Japan), or hydrophilic polyvinylidene difluoride (PVDF) (pore size 5.0 μm) (Millipore, Tokyo, Japan), respectively. The DNA sample was stored at -60 °C until the real-time PCR assay was performed.

Based on the results of the extraction efficiency tests on the various tests, a standard curves consisting of eight-fold serial dilutions (10^4 to 1 cells) of cultured cells were prepared. Each number of the cells was collected on the Nuclepore filter (pore size 3.0 μm) (Whatman, Maidstone, UK) by filtration. The DNA extraction was performed by the TE boiling method, and the real-time PCR assay was carried out in triplicate. The standard curve was constructed based on the correlation between the threshold cycle (Ct value) and the number of cells.

A major concern when designing a real-time PCR assay for HABs is whether other co-occurring microalgae adversely affect the amplification efficiency either by introducing inhibitors or due to cross-reactivity problems. This possibility was explored in an examination where *H. circularisquama* cells (10^4 to 1 cells) were filtered on the Nuclepore filters (pore size 3.0 μm) at 20 cm Hg with 10^5 cells each of *H. triquetra, H. akashiwo, K. mikimotoi* and *Skeletonema* sp. which are frequently co-dominated in western coastal areas of Japan. A previous study also showed that the primers and probe used in this study are species-specific and do not react DNA from *H. triquetra, H. akashiwo* or *K. mikimotoi* (Kamikawa et al., 2006). The DNA extraction was performed by the TE boiling method, and the real-time PCR assay was carried out as follows in triplicate. Obtained Ct values at each number of cells were compared with those of the control experiment where only *H. circularisquama* was used.

4.3 Real-time PCR

The primer set and probe used in this study were based on unique species-specific DNA sites identified by aligning the D1/D2 LSU rDNA sequence of *H. circularisquama* (DDBJ/EMBL/GenBank accession number AB049709) with the correponding dinoflagellate sequences in GenBank. Primers specific to *H. circularisquama* were HcirF (5′-GTTTGCCTATGGGTGAGC-3′) and HcirR (5′-CATTGTGTCAGGGAGGAG-3′) and the probe was HcirTaqMan (5′-FAM-CACCACAAGGTCATGAGGACACA-TAMRA-3′) that was labeled at the 5′-end with FAM (carboxyfluorescein) and the 3′-end with TAMRA (carboxytetramethylrhodamine) (Kamikawa et al., 2006).

Thermal cycling was performed with a Rotor-Gene 3000 (Corbett Research, Mortlake, Australia) in 200-μL PCR tubes of commodity type. PCR was carried out in 25-μL volumes comprising 1×PCR EX Taq buffer (containing 20mM Mg2+), 200 μM dATP, 200 μM dTTP, 200 μM dGTP, 200 μM dCTP, 0.3 μM forward and reverse primers, 0.4 μM fluorogenic probe, and 1.25 U of Taq DNA polymerase (Takara EX TaqTM, TaKaRa Bio Inc., Shiga, Japan). The PCR conditions were as follows according to Kamikawa et al. (2006): one heating cycle at 95 °C for 2 min, followed by 45 cycles at 95 °C for 10 sec and 54 °C for 30 sec. The Ct value was calculated as a cycle number that an amplification curve reached at the most suitable threshold value.

5. Results and discussion

5.1 Development of a DNA extraction method

In order to examine the most efficient method for DNA extraction, three kinds of DNA extraction methods were subjected to *H. circularisquama* cells (100 cells and 1 cell) trapped on the filter. Figure 2 shows obtained Ct values for one and 100 cells by real-time PCR assay with each DNA extraction method. For 100 cells, the DNA extracted with the TE boiling method was as efficient as with the CTAB method and the proteinase K method (t-test, df = 4, $p > 0.05$). For 1 cell, the DNA extraction efficiency with the TE boiling method was higher than that with the modified CTAB method (t-test, df = 4, $p < 0.05$) and similar to that of the proteinase K method (t-test, df = 4, $p > 0.05$). Thus, we can consider that the three methods are similarly efficient for *H. circularisquama* cells with high density. Given the importance of detection the HAB species at low density, the TE boiling method appeared to be the most useful technique for monitoring *H. circularisquama* by real-time PCR assay. In addition to its higher detection efficiency, the TE boiling method is more suitable in simplicity, ease of execution, lower cost, and shorter execution time than the other two methods.

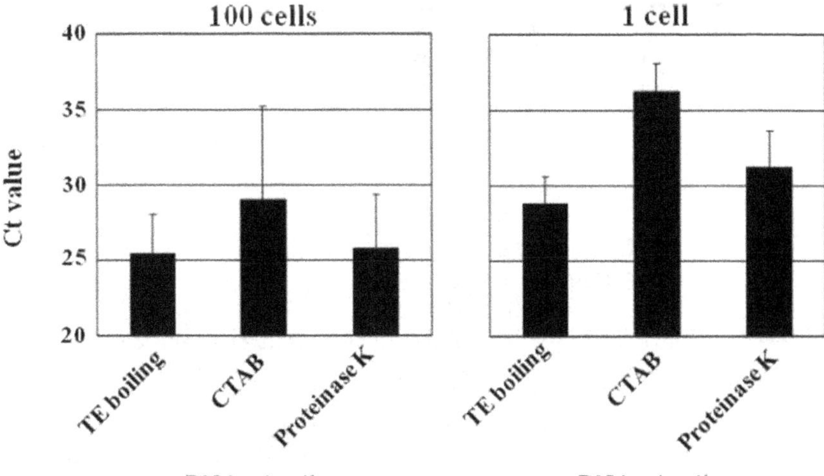

Fig. 2. Comparison of Ct (Threshold cycle) values obtained with three DNA extraction methods. TE boiling method, CTAB method and proteinase K and SDS method were subjected to 100 cells (left) and one cell (right) of *H. circularisquama* on Nucleporepolycarbonate membrane filters. Ct values were obtained by using the real-time PCR assay in triplicate. The bars show the standard deviations.

In order to select the filter which would yield the highest and most consistent recovery of DNA, samples containing either 1 or 100 cells of *H. circularisquama* cells were filtered onto six different types of filter. DNA was then extracted using the TE boiling method and subjected to real-time PCR-amplification. In the case of 100 cells, the real-time PCR assay successfully amplified *H. circularisquama* DNA from all the filters with the exception of the polytetra fluoroethylene (PTFE) membrane filter (Fig. 3). In contrast, for the 1 cell samples, the qPCR assay failed to reliably amplify the DNA from all the filters with the exception of the Nuclepore polycarbonate membrane filter, which could be detected in triplicate. Only one of three DNA samples extracted from either the cellulose acetate or hydrophilic polyvinylidine difluoride (PVDF) membrane filters were detected. Therefore, we concluded that the best filter for extraction and detection of *H. circularisquama* was Nuclepore polycarbonate membrane filter.

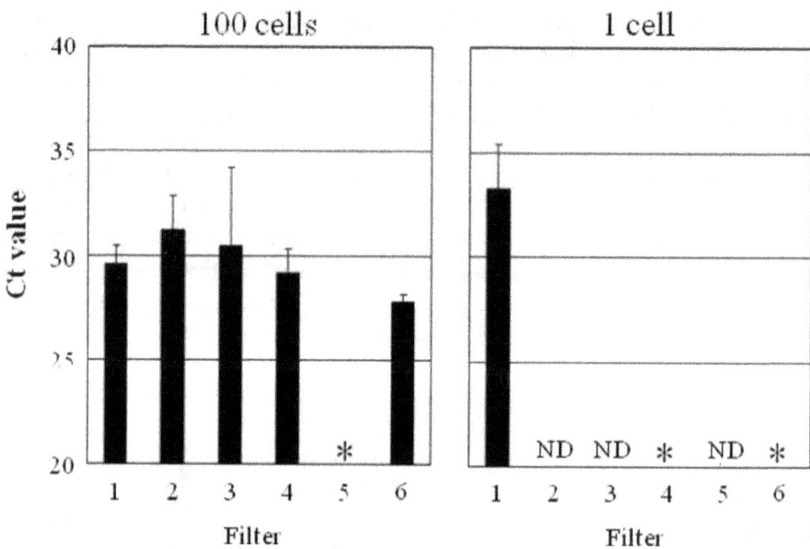

Fig. 3. Ct values obtained with TE boiling DNA extraction method for 100 cells (left) and 1 cell (right) collected on six different filters using the real-time PCR assay. 1, Polycarbonate filter; 2, Glass fiber filter; 3, Cellulose mixed ester filter; 4, Cellulose acetate membrane filter; 5, PTFE membrane filter; 6, Hydrophilic PVDF filter. Asterisks (*) indicate that Ct value could be obtained from only one of three filters. ND means that Ct value could not be obtained from all three filters. The bars show the standard deviations.

In a previous study (Kamikawa et al. 2006), *H. circularisquama* cells were concentrated by natural gravity filtration. However, it takes whole day for the concentration by natural filtration. It is not feasible for the routine works for monitoring natural populations. In addition, concentrating cells by filtration is not feasible for *H. circularisquama*, because most cells are attached and trapped on to the surface of any filters examined (Shiraishi et al., 2007).

Otherwise, more amount of seawater for monitoring is more suitable for detecting cell during period of low cell density, indicating that concentration of cells from seawater is important for accurate and sensitive detection.

The experiments above demonstrates that cultured cells of *H. circularisquama* can be quantitatively recovered and amplified from a single cell gently filtered onto Nucleopore filters (pore size 3.0 μm) and extracted using the boiling TE method (Fig. 4).

5.2 Validity of the real-time PCR assay

Serial dilutions of vegetative cells (1-10^4 cells) of *H. circularisquama* on Nuclepore filters were prepared, and then, the DNA was extracted by the TE boiling method. The real-time PCR assay was performed with the DNA samples in triplicates. The standard curve was constructed based on the mean Ct values and the number of *H. circularisquama* cells (Fig. 4). The obtained relationship between Ct values and the number of cells in logarithmic scale was linear, and the correlation coefficient was significantly high (r^2 = 0.997), indicating that the simple real-time PCR protocol can quantitatively detect *H. circularisquama* even from one cell.

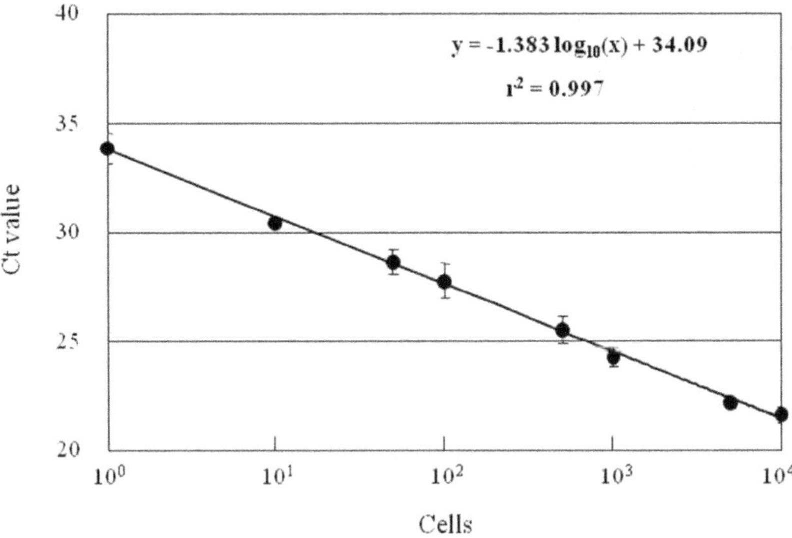

Fig. 4. Real-time PCR assay using eight-fold serial dilutions (10^4 to 1 cell). The result for each cell number was represented by each symbol shown in the figure.

In order to examine the effects of the existence of other microalgae on the DNA extraction or subsequent PCR-based quantification, *H. circularisquama* cells (10^4 to 1 cells) were collected together with 10^5 cells of *H. triquetra, H. akashiwo, K. mikimotoi* and *Skeletonema* sp. on the Nuclepore membrane filters by filtration. The DNA extraction and the real-time PCR assay were carried out as described above. The standard curve was constructed based on the mean Ct values and the number of cells (Fig. 5a). At each number of *H. circularisquama* examined, there was no significant difference between the Ct value obtained from *H. circularisquama* cells in spite of presence or absence of the other microalgae (t-test, df = 4, $p >$ 0.05, Fig. 5b). The correlation between Ct values and the number of cells in logarithmic scale was linear, and the correlation coefficient was extremely significant (r^2 = 0.991, Fig. 5a). It was confirmed that the DNA extraction and subsequent PCR-based quantification of *H. circularisquama* cells were not inhibited even when other microalgae such as *H. triquetra, H. akashiwo, K. mikimotoi* and *Skeletonema* sp. coexist with *H. circularisquama*.

The constructed standard curve showed linearity (Fig. 4), and the protocol including concentration of cells, DNA extraction, and the real-time PCR was not inhibited by the existence of other microalgae even at 10^5 cells of *H. triquetra, H. akashiwo, K. mikimotoi* and *Skeletonema* sp. (Fig. 5b). It was clearly demonstrated that the presence of closely related species (e.g., *H. triquetra*) and/or many other common red tide species did not affect the efficiency of DNA extraction and subsequent PCR-based quantification of *H. circularisquama* cells.

When there are *H. circularisquama* cells in addition to much higher abundance of similar sized phytoplankton in the field, the detection of *H. circularisquama* cells by the canonical IFAT method with epifluorescent signal is obscured by the presence of numerous other microalgal cells. Similarly, an underestimation of cell abundance can be occurred when a large number of particles such as detritus not only inhibit epifluorescence microscopy observation but also blocks the antibody reaction trapped within the detritus, indicating that the IFAT method is difficult to be applied to sediments and detritus-rich samples. The real-time PCR assay described in this study appears to be more feasible and practical for environmental samples than the IFAT method.

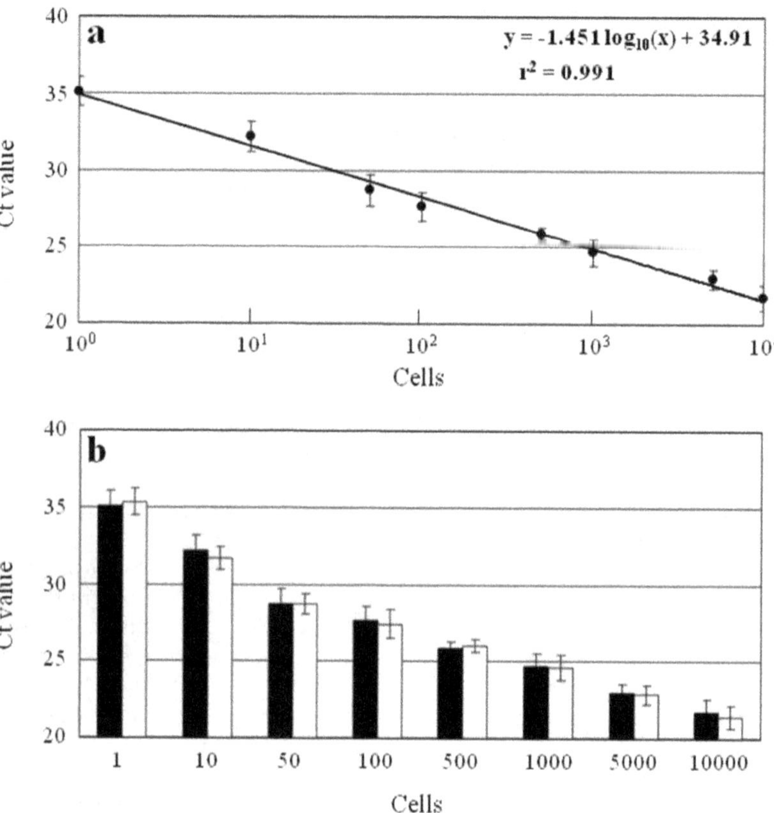

Fig. 5. Relationship between Ct values and the log number of cells. **a**. Standard curve for *H. circularisquama* cells constructed from DNA that was extracted from *H. circularisquama* cells plus several microalgae. **b**. Comparison of detection and quantification efficiency between DNA that was extracted from *H. circularisquama* cells (closed bars) and from *H. circularisquama* cells plus several microalgae (open bars). The bars show standard deviations.

6. Application to environmental samples

The procedure described above was applied to environmental samples in order to monitor successively *H. circularisquama* cells in addition to IFAT assay (Shiraishi et al. 2009.). The cell densities obtained by the real-time PCR assay were almost identical to the results obtained by the IFAT method. Hence, it was clearly demonstrated that *H. circularisquama* could be quantified by this simple real-time PCR assay as sensitively and precisely as the IFAT method in the field. It is notable that the detection limit of the real-time PCR assay was 1 cell/L: The most sensitive level currently with real-time PCR assay (Shiraishi et al. 2009.). It should be also mentioned that the real-time PCR assay sometimes reacted to some environmental samples which the IFAT assay did not (Shiraishi et al. 2009.). This incongruence can be explained by that the real-time PCR assay is more sensitive than the IFAT assay. Otherwise, the real-time PCR assay might react to cell-free DNA derived from broken, dead cells of *H. circularisquama*. Since we have no idea which is true, it is better to use both methods for monitoring *H. circularisquama* cells in order to grasp precise dynamics of the HAB species, escaping both underestimation and overestimation.

In this chapter, it has been demonstrated that the real-time PCR assay can be applied to monitoring various HABs in field waters. If other HABs can be quantified by the same manner to present method with slight modification, those microalgae would be easily monitored with the similar procedures of the DNA extraction at the same time. The conventional methods for monitoring HABs with optical microscopy might be replaced by the simple real-time PCR assay in the near future, when the costs of machines and reagents are lowered to become reasonable.

In addition to seawater samples, real-time PCR assay has been applied for the detection of the cysts of the toxic *Alexandrium* species from marine sediments (Kamikawa et al., 2005, 2007, Erdner et al. 2011). Furthermore, the PCR method was also used for the detection of the cells of *Alexandrium* species in the tissue of mussels (Galluzzi et al., 2005) in order to investigate the possibility that the HAB cells are propagated to other areas by transport of bivalves. When *H. circularisquama* forms temporary cysts in water columns, those temporary cysts possibly sink down and survive some periods at the surface of the sea bottom. There are some reports that *H. circularisquama* could proliferate in water columns of a new area after the transportation of bivalves which accompany temporary cysts (Honjo et al., 1998; Honjo and Imada, 1999; Imada et al., 2001). Given the possibility of the temporary cysts as a seed-population, the detection of *H. circularisquama* cells is an urgent need from sediments, and tissues and fecal pellets of bivalves.

7. Acknowledgment

We would like to thank Drs. T. Uchida, M. Yamaguchi, and Y. Matsuyama (the National Research Institute of Fisheries and Environment of Inland Sea, Fisheries Research Agency), and G. Nishitani (Tohoku University) for their kind donation of the culture strains of *H. circularisquama, K. mikimotoi* and *Skeletonema* sp. for this study. RK is a research fellow supported by the Japan Society for Promotion of Sciences (no. 210528). This work wa supported in part by a grant from the Fishery Agency of Japan.

8. References

Adachi, M.; Sako, Y. & Ishida, Y. (1994). Restriction fragment length polymorphism of ribosomal DNA internal transcribed spacer and 5.8S regions in Japanese *Alexandrium* species (Dinophyceae). *Journal of Phycology*, Vol.30, pp.857-863.

Anderson, D.M. (1994). Red tides. *Scientific American*, Vol.271, pp.52-58.

Bowers, H.A.; Tengs, T., Glasgow, H.B.Jr, Burkholder, J.M., Rublee, P.A. & Oldach, D.W. (2000). Development of real-time PCR assays for rapid detection of Pfiesteria piscicida and related dinoflagellates. *Applied and Environmental Microbiology*, Vol.66, pp.4641-4648

Bowers, H.A.; Tomas, C., Tengs, T., Kempton, J.W., Lewitus, A.J. & Oldach, D.W. (2006). Raphidophyceae [Chadefaud ex Silva] systematic and rapid identification: sequence analyses and real-time PCR assays. *Journal of Phycology*, Vol.42, pp.1333-1348.

Chen, L.C.M.; Edelstein, T. & McLachlan, J. (1969). *Bonnemaisonia hamifera* Hariot in nature and in culture. *Journal of Phycology*, Vol.5, pp.211-220.

Dyhrman, S.T.; Erdner, D., Du, J.L., Galac, M. & Anderson, D.M. (2006). Molecular quantification of toxic *Alexandrium fundyense* in the gulf of Maine using real-time PCR. *Harmful Algae*, Vol.5, pp.242-250.

Erdner, D.; Percy, L., Keafer, B., Lewis, J. & Anderson, D.M. (2011). A quantitative real-time PCR assay for the identification and enumeration of *Alexandrium* cysts in marine sediments. *Deep Sea Research Part II: Topical Studies in Oceanography*, Vol.57, pp.279-287.

Galluzzi, L.; Penna, A., Bertozzini, E., Vila, M., Garcés, E. & Magnani, M. (2004). Development of a real-time PCR assay for rapid detection and quantification of *Alexandrium minutum* (a Dinoflagellate). *Applied and Environmental Microbiology*, Vol.70, pp.1199-1206.

Galluzzi, L.; Penna, A., Bertozzini, E., Giacobbe, M.G., Vila, M., Garcés, E., Prioli, S. & Magnani, M. (2005). Development of a quantitative PCR method for the *Alexandrium* spp. (Dinophyceae) detection in contaminated mussels (*Mytilus galloprovincialis*). *Harmful Algae*, Vol.4, pp.973-983.

Godhe, A.; Rehnstam-Holm, A.S., Karunasagar, I. & Karunasagar, I. (2002). PCR detection of dinoflagellate cysts in field sediment samples from tropic and temperate environments. *Harmful Algae*, Vol.1, pp.361-373.

Gray, M.; Wawrik, B., Paul, J. & Casper, E. (2003). Molecular detection and quantitation of the red tide dinoflagellate *Karenia brevis* in the marine environment. *Applied and Environmental Microbiology*, Vol.69, pp.5726-5730.

Hallegraeff, G.M. (1993). A review of harmful algal blooms and their apparent global increase. *Phycologia*, Vol.32, pp.79-99.

Hallegraeff, G.M. (1995). Harmful algal blooms: a global overview. In: G.M. Hallegraeff, D.M. Anderson and A.D. Cembella, Editors, Manual on Harmful Marine Microalgae, IOC Manuals and Guides. UNESCO. Vol.33, pp.1–22.

Handy, S.M.; Coyne, K.J., Portune, K.J., Demir, E., Doblin, M.A., Hare, C.E., Cary, S.C. & Hutchins, D.A. (2005). Evaluating vertical migration behavior of harmful raphidophytes in the Delaware Inland Bays utilizing quantitative real-time PCR. *Aquatic Microbial Ecology*, Vol.40, pp.121-132.

Heid, C.A.; Stevens, J., Livak, K.J. & Williams, P.M. (1996). Real time quantitative PCR. *Genome Research*, Vol.6, pp.986-994.

Holland, P.M.; Abramson, R.D., Watson, R. & Gelfand D.H. (1991). Detection of specific polymerase chain reaction product by utilizing the 5'-3' exonuclease activity of Thermus aquaticus DNA polymerase. *Proceedings of the National Academy of Sciences of the United States of America*, Vol.88, pp.7276-7280.

Honjo, T.; Imada, N., Ohshima, Y., Maema, Y., Nagai, K., Matsuyama, Y. & Uchida, T. (1998). Potential transfer of Heterocapsa circularisquama with pearl oyster consignments. In: Reguera, B., Blanco, J., Fernandez, M.L., Wyatt, T. (Eds), Harmful Algae. Xunta de Galicia and IOC of UNESCO, Santiago de Compostela, pp.224-226.

Honjo, T. & Imada, N. (1999). Future attention-Spread of *Heterocapsa circularisquama* red tides and its preventive measure. *Bulltin of the Plankton Society of Japan*, Vo.46, pp.180-181 (in Japanese).

Horiguchi, T. (1995). *Heterocapsa circularisquama* sp. nov. (Peridiniales, Dinophyceae): a new marine dinoflagellate causing mass mortality of bivalves in Japan. *Phycological Research*, Vol.43, pp.129-136.

Hosoi-Tanabe, S. & Sako, Y. (2005a). Rapid detection of natural cells of *Alexandrium tamarense* and *A. catenella* (Dinophyceae) by fluorescence *in situ* hybridization. *Harmful Algae*, Vol.4, pp.319-328.

Hosoi-Tanabe, S. & Sako, Y. (2005b). Species-specific detection and quantification of toxic marine dinoflagellates *Alexandrium tamarense* and *A. catenella* by real-time PCR assay. *Marine Biotechnology*, Vol.7, pp.506-514.

Imada, N.; Honjo, T., Shibata, H., Oshima, Y., Nagai, K., Matsuyama, Y. & Uchida, T. (2001). The quantities of *Heterocapsa circularisquama* cells transferred with shellfish consignments and the possibility of its establishment in new areas. In: Hallegraeff, G.M., Blackburn, S.I., Bolch, C.J., Lewis, R.J. (Eds), Harmful Algal Blooms 2000. IOC of UNESCO, Paris, pp.474-476.

Imai, I.; Itakura, S., Matsuyama, Y. & Yamaguchi, M. (1996). Selenium requirement for growth of a novel red tide flagellate *Chattonella verruculosa* (Raphidophyceae) in culture. *Fisheries Science*, Vol.62, pp.834-835.

Imai, I. (2000). Current problems in classification and identification of marine raphidoflagellates (raphidophycean flagellates): from the view point of ecological study. *Bulltin of the Plankton Society of Japan*, Vol.47, pp.55–64 (in Japanese).

Imai, I.; Yamaguchi, M. & Hori, Y. (2006). Eutrophication and occurrences of harmful algal blooms in the Seto Inland Sea, Japan. *Plankton and Benthos Ressearch*, Vol.1, pp.71-84.

Iwataki, M.; Hansen, G., Sawaguchi, T., Hiroishi, S. & Fukuyo, Y. (2004). Investigations of body scales in twelve *Heterocapsa* species (Peridiniales, Dinophyceae), including a new species *H. pseudotriquetra* sp. nov. *Phycologia* Vol.43, pp.394-403.

Kamikawa, R.; Hosoi-Tanabe, S., Nagai, S., Itakura, S. & Sako, Y. (2005). Development of a quantification assay for the cysts of the toxic dinoflagellate *Alexandrium tamarense* using real-time polymerase chain reaction. *Fisheries Science*, Vol.71, pp.987-991.

Kamikawa, R.; Asai, J., Miyahara, T., Murata, K., Oyama, K., Yoshimatsu, S., Yoshida, T. & Sako, Y. (2006). Application of a real-time PCR assay to a comprehensive method of monitoring harmful algae. *Microbes and Environments*, Vol.21, pp.163-173.

Kamikawa, R.; Nagai, S., Hosoi-Tanabe, S., Itakura, S., Yamaguchi, M., Uchida, Y., Baba, T. & Sako, Y. (2007). Application of real-time PCR assay for detection and quantification of *Alexandrium tamarense* and *Alexandrium catenella* cysts from marine sediments. *Harmful Algae*, Vol.6, pp.413-420.

Matsuyama, Y.; Uchida, T., Nagai, K., Ishimura, M., Nishimura, A., Yamaguchi, M. & Honjo, T. (1996). Biological and environmental aspects of noxious dinoflagellate red tides by *Heterocapsa circularisquama* in the west Japan. In: Yasumoto, T., Oshima, Y., Fukuyo, Y. (Eds), Harmful and Toxic Algal Blooms. UNESCO, Paris, pp.247-250.

Matsuyama, Y.; Kimura, A., Fujii, H., Takayama, H. & Uchida, T. (1997). Occurrence of a *Heterocapsa circularisquama* red tide and subsequent damages to shellfish in western Hiroshima Bay, Seto Inland Sea, Japan in 1995. *Bulletin of the Nansei National Fisheries Research Institute*, Vol.30, pp.189-207 (in Japanese, with English abstract).

Matsuyama, Y. (1999). Harmful effect of dinoflagellate *Heterocapsa circularisquama* on shellfish aquaculture in Japan. *Japan Agricultural Research Quarterly*, Vol.33, pp.283-293.

Matsuyama, Y.; Uchida, T., Honjo, T. & Shumway, S.E. (2001). Impacts of the harmful dinoflagellate, *Heterocapsa circularisquama*, on shellfish aquaculture in Japan. *Journal of Shellfish Research*, Vol.20, pp.1269-1272.

Nagai, K.; Matsuyama, Y., Uchida, T., Yamaguchi, M., Ishimura, M., Nishimura, A., Akamatsu, S. & Honjo, T. (1996). Toxicity and LD50 levels of the red tide dinoflagellate *Heterocapsa circularisquama* on juvenile pearl oysters. *Aquaculture*, Vol.144, pp.149-154.

Nagai, K.; Matsuyama, Y., Uchida, T., Akamatsu, S. & Honjo, T. (2000). Effect of a natural population of the harmful dinoflagellate *Heterocapsa circularisquama* on the survival of the pearl oyster *Pinctada fucata*. *Fisheries Science*, Vol.66, pp.995-997.

Nakanishi, K.; Onaka, S., Kobayashi, T. & Masuda, K. (1999). Bloom dynamics of *Heterocapsa circularisquama* in Ago Bay, Japan. *Bulltin of the Plankton Society of Japan*, Vol.46, pp.161-164 (in Japanese).

Sako, Y.; Hosoi-Tanabe, S. & Uchida, A. (2004). Fluorescence in situ hybridization using rRNA-targeted probes for simple and rapid identification of the toxic dinoflagellates *Alexandrium tamarense* and *Alexandrium catenella*. *Journal of Phycology*, Vol.40, pp.598-605.

Shiraishi, T.; Hiroishi, S., Nagai, K., Go, J., Yamamoto, T. & Imai, I. (2007). Seasonal distribution of the shellfish-killing dinoflagellate *Heterocapsa circularisquama* in Ago Bay monitored by an indirect fluorescent antibody technique using monoclonal antibodies. *Plankton Benthos Research*, Vol.2, pp.49-62.

Shiraishi, T.; Hiroishi, S., Kamikawa, R., Sako, Y., Taino, S., Ishikawa, T., Hayashi, Y. & Imai, I. (2009). Population dynamics of the shellfish-killing dinoflagellate *Heterocapsa circularisquama* monitored by an indirect fluorescent antibody technique and a real-time PCR assay in Uranouchi Inlet, Kochi Prefecture, Japan. In Proceedings of 5th World Fisheries Congress, 6c_1006_200, TerraPub, Tokyo.

Shiraishi, T.; Hiroishi, S., Taino, S., Ishikawa, T., Hayashi, Y., Sakamoto, S., Yamaguchi, M. & Imai, I. (2008). Identification of overwintering vegetative cells of the bivalve-killing dinoflagellate *Heterocapsa circularisquama* in Uranouchi Inlet, Kochi Prefecture, Japan. *Fisheries Science*, Vol.74, pp.128-136.

Smayda, T.J. (1997). Bloom dynamics: physiology, behaviour, trophic effects. *Limnology and Oceanography*, Vol.42, pp.1132-1136.

Tamai, K. (1999). Current status of outbreaks and fisheries damages due to Heterocapsa circularisquama. *Bulltin of the Plankton Society of Japan*, Vol.46, pp.153-154 (in Japanese).

Yamaguchi, M.; Itakura, S., Imai, I. & Ishida, Y. (1995). A rapid and precise technique for enumeration of resting cysts of *Alexandrium* spp. (Dinophyceae) in natural sediments. *Phycologia*, Vo.34, pp.207–214.

Zhang, H. & Lin, S. (2005). Development of a cob-18S rRNA gene real-time PCR assay for quantifying *Pfiesteria shumwayae* in the natural environment. *Applied and Environmental Microbiology*, Vo.71, pp. 7053-7063.

Zhou, Z.; Miwa, M. & Hogetsu, T. (1999). Analysis of genetic structure of a *Suillus grevillei* population in a *Larix kaempferi* stand by polymorphism of inter-simple sequence repeat (ISSR). *New Phytologists*, Vo.144, pp. 55-63.

2

Species Identification of Food Spoilage and Pathogenic Bacteria by MALDI-TOF Mass Fingerprinting

Karola Böhme[1], Inmaculada C. Fernández-No[1], Jorge Barros-Velázquez[1], Jose M. Gallardo[2], Benito Cañas[3] and Pilar Calo-Mata[1]
[1]*Department of Analytical Chemistry, Nutrition and Food Science, School of Veterinary Sciences, University of Santiago de Compostela, Lugo,*
[2]*Department of Food Technology, Institute for Marine Research (IIM-CSIC), Vigo,*
[3]*Department of Analytical Chemistry, University Complutense of Madrid, Madrid,*
Spain

1. Introduction

Food quality and safety is an increasingly important public health issue. Nowadays, the topics "food quality" and "food safety" are very close and two important issues in the food sector, due to the globalization of the food supply and the increased complexity of the food chain. The consumers need to purchase safe products that do not involve any kind of risk for health. On one hand, the aim of the "food safety" is to avoid health hazards for the consumer: microbiological hazards, pesticide residues, misuse of food additives and contaminants, such as chemicals, biological toxins and adulteration. On the other hand, "food quality" includes all attributes that influence the value of a product for the consumer; this includes negative attributes such as spoilage, contamination with filth, discoloration, off-odors and positive attributes such as the origin, color, flavor, texture and processing method of the food (FAO, 2003).

The contamination of food products with microorganisms presents a problem of global concern, since the growth and metabolism of microorganisms can cause serious foodborne intoxications and a rapid spoilage of the food products. Thus, the acceptance and safety of a food product for the consumers depends in great part on the presence and nature of microorganisms. Besides molds and yeasts, bacteria are the principle responsible for various types of food spoilage and foodborne intoxications (Blackburn, 2006). It has to be mentioned that a food product naturally contains an indigenous microbiota that can include spoilage and/or pathogenic bacterial species. Depending on the preservation method these species can proliferate and adulterate the product. However, most bacterial contamination occurs during processing and manipulation of the food products.

1.1 Spoilage bacteria

Spoilage bacteria are microorganisms that cause the deterioration of food and develop unpleasant odors, tastes, and textures. A spoiled food has lost the original nutritional value, texture or flavor and can become harmful to people and unsuitable to eat. The microbial spoilage of food products constitutes an important economic problem, as it results in high economic losses for the food industry, especially under incorrect refrigeration conditions. Thus, spoilage bacteria are able to grow in large number in food, decompose the food and cause changes in the taste/smell, which affect the quality of the products. Spoilage bacteria normally do not cause illness; however, when consumed in high concentration, they can cause gastrointestinal disturbance (Blackburn, 2006). There are different bacterial species that can cause spoilage in food products and the spoilage microbiota depends in great part on the processing and preservation method. Storage temperature also plays a key role in the growth of undesirable microbiota in food. Thus, fresh foodstuffs such as fish and meat, stored at refrigeration temperatures can result in the growth of *Pseudomonas* spp., including spoilage species, such as *P. fragi* and *P. putida*. A light preservation and change in the atmosphere, e.g. by vacuum-packaging, may inhibit these bacterial species and favour the growth of other species, such as lactic acid bacteria (LAB), *Enterobacteriaceae, Bacillus* spp. and *Clostridium* spp. These last two genera are able to produce spores that can survive heat treatments and germinate after a pasteurization process, being an important issue in food safety. Spoilage species may be more food-specific and, thus, *Erwinia* spp. has been reported in products of vegetal origin. Otherwise, seafood products are commonly spoiled by species such as *Shewanella* spp. or *Photobacterium* spp. In general, bacteria can spoil different foods depending on the physical-chemical preservation profile (Gram et al., 2002).

1.2 Pathogenic bacteria

Foodborne diseases are caused by agents that enter the body through the ingestion of food. Food can transmit disease from person to person, as well as serve as a growth medium for bacteria that can cause food poisoning. The global incidence of foodborne diseases is difficult to estimate, but it has been reported that in 2005 alone 1.8 million people died from diarrheal diseases. A great proportion of these cases can be attributed to the consumption of contaminated food and water. In industrialized countries, the percentage of the population suffering from foodborne diseases each year has been reported to be up to 30% (WHO, 2007). Pathogenic bacteria often do not change the color, odor, taste or texture of a food product, being hard to recognize if the product is contaminated. Food-borne infection is caused by bacteria in food. If bacteria become numerous and the food is eaten, bacteria may continue to grow in intestines and cause illness. Food intoxication results from consumption of toxins (or poisons) produced in food as a by-product of bacterial growth and multiplication in food. In this case the toxins and not bacteria cause illness. Toxins may not alter the appearance, odor or flavor of food. Common bacteria that produce toxins include *Staphylococcus aureus* and *Clostridium botulinum*. In some cases, such as *Clostridium perfringens*, illness is caused by toxins released in the gut, when large numbers of vegetative cells are eaten. Bacterial food poisoning is commonly caused by bacterial pathogenic species such as *Escherichia coli, Salmonella* spp., *Listeria monocytogenes, S. aureus, Bacillus cereus, C. perfringens, Campylobacter* spp., *Shigella* spp., *Streptococcus* spp., *Vibrio cholerae,* including O1 and non-O1, *Vibrio parahaemolyticus, Vibrio vulnificus, Yersinia enterocolitica* and *Yersinia pseudotuberculosis.* Emerging foodborne pathogens may refer to new pathogens, pathogens that emerge due to changing ecology or changing technology that connects a potential pathogen with the food chain or emerge de novo by transfer of mobile virulence factors (Tauxe, 2002). Emerging foodborne pathogens include *E. coli* O157:H7, *Aeromonas hydrophila, Aeromonas caviae, Aeromonas sobria, Mycobacterium* spp., vancomycin-resistant enterococci, non-gastric *Helicobacter* spp., *Enterobacter sakazakii,* non-jejuni/coli species of *Campylobacter,* and non-O157 Shiga toxin-producing *E. coli.*

2. Bacterial identification methods

In order to control and minimize the microbiological hazard of foodborne pathogens, as well as to predict and enhance shelf-life of food products, pathogenic and spoilage bacteria need to be identified in a rapid and unequivocal way. Several methods have been designed to achieve bacterial identification.

2.1 Bacterial identification by classic methods

Traditionally, bacterial species have been identified by classic tools relying on culturing processes coupled to morphological, physiological, and biochemical characterization. Phenotypic identification is based on direct comparison of phenotypic characteristics of unknown bacteria with those of type cultures. The reliability of this kind of identification is in direct proportion to the number of similar phenotypic characteristics.

When classifying microorganisms, all known characteristics are taken into account. However, certain characteristics are selected and used for the purpose of identification. Primary identification usually involves a few simple assays such as colony morphology, Gram staining, growth conditions, catalase and oxidase tests (Duerden et al., 1998). Testing the requirements for growth includes the presence or absence of oxygen and the growth ability on different culture media. For a better approximation of bacterial identification, other laborious techniques are employed, such as microscopic observation, type of hemolysis, and biochemical arrays like tests for aminopeptidase, urease, indol, oxido-fermentation, coagulase test, analysis of resistance to different substances, etc. To make these analyses faster and less laborious, biochemical assays with multitest galleries are applied. With these arrays inoculation, incubation and lecture can be carried out on a minimal space and the whole process can be automated. Using these tests it is usually possible to characterize the bacterial genus and even species to that an unknown strain more likely belongs to. Conventional identification methods are widely used despite some disadvantages. Apart from being slow and laborious, they can be used only for organisms that can be cultivated in vitro and furthermore, some strains exhibit unique biochemical characteristics that do not fit into patterns that have been used as a characteristic of any known genus and species.

In the last decades, the progress of microbiological identification turned to more rapid and sensitive methods, including antibody-based assays and DNA-based methods, together with important advances in bioinformatic tools. Thus, some methodologies such as ELISA or PCR already became classic. PCR coupled to sequencing tools has provided a big amount of information that has been deposited in public databases and is freely available. Recently, the development of rapid and high sensitive techniques, such as real-time PCR (RTi-PCR), DNA microarrays and biosensors, provoked the replacement of traditional culturing methods in the field of bacterial identification in clinical diagnostics, as well as in the food sector (Feng, 2007; Mohania et al., 2008). Furthermore, Fourier transform infrared spectroscopy (FT-IR) has been described as a new method for rapid and reliable bacterial identification (Sandt et al., 2006). At the same time, proteomic tools, such as mass spectrometry were introduced for the identification of microorganisms (Klaenhammer et al., 2007).

2.2 Bacterial identification by DNA-based methods

Nowadays, in the field of bacterial diagnostics, traditional culturing methods have been replaced by molecular techniques based on the analysis of DNA, being much faster, more sensitive and accurate. However, they usually have a superior cost due to that they require specific industrial equipment and more qualified personal than conventional techniques.

The sequencing of the 16S rRNA gene is a common tool for bacterial identification. Universal primers are designed and bind to conserved regions to amplify variable regions. The amplification is carried out by means of PCR (Polymerase Chain Reaction). Sequences of the 16S rRNA gene have been determined for an extremely large number of species and are accessible on huge DNA databases, such as the GenBank of the National Center for Biotechnology Information (NCBI). For bacterial identification the 16S rRNA sequence of an unknown strain is compared to the database of published sequences and the most similar bacterial strains are determined using the common bioinformatic tool BLAST.

Other DNA genes have also been used for the study of phylogenetic relationships, such as the 23S rRNA, the intergenic spacer region of 16S-23S (ITS), rpoB (subunit β of RNA polymerase) and gyrB (subunit β of the DNA girasa). However, for bacterial identification these genes are not yet applied, due to the lack of amply databases as reference.

More recently, RTi-PCR has been introduced for the detection and quantification of major foodborne spoilage and/or pathogenic bacteria. As a consequence, a large number of RTi-PCR procedures are currently available for the specific detection and quantification of foodborne bacteria, such as *Salmonella spp., L. monocytogenes, S. aures* or *Leuconostoc mesenteroides* (Hein et al., 2001; Malorny et al., 2004; Elizaquível et al., 2008).

The state-of-the-art DNA-based technique is DNA microarray that consists of gene arrays that can hybridize multiple DNA targets simultaneously, and thus, have enormous potential for detection and identification of pathogens. Thanks to the increase in the complete microbial genome sequences, DNA microarrays are becoming a common tool in many areas of microbial research in the field of bacterial identification in clinical diagnostics, as well as in the food sector (Severgnini et al., 2011). Microarray is a powerful, sensitive and specific technology that allows an accurate identification based on single target detection and can determine subtle differences in the genome.

2.3 Mass spectrometry for bacterial identification

In the past few years, proteomic tools, such as mass spectrometry, were introduced for microbial identification (Klaenhammer et al., 2007). According to the ionization source, mass spectrometry can be divided in two techniques electrospray ionization (ESI) and matrix-assisted laser desorption/ ionization (MALDI). The mass analyzer uses an electric or magnetic field in order to accelerate the ions and produce their separation, by means of the ratio mass/charge (m/z). In the field of bacterial identification, mass spectrometric methods have a high potential, due to the ability to detect and identify bacterial proteins (van Baar, 2000).

Matrix-assisted laser desorption ionization-time of flight mass spectrometry (MALDI-TOF MS) proved to be a technique with high potential for microorganism identification due to its rapidness, reduced cost and minimal sample preparation compared to conventional biochemical and molecular techniques (Seng et al., 2009; Bizzini et al., 2010; Giebel et al., 2010). With MALDI-TOF MS intact bacterial cells can be analyzed in a rapid way, obtaining high specific spectral profiles in the low-mass range of 1500 - 20000 Da. The soft ionization technique allows the analysis of intact high mass molecules, such as proteins. Studies, in that bacterial cells were treated with trypsin and lysozyme before analysis by MALDI-TOF MS showed that the spectral patterns detected by this method are generally attributed to intracellular proteins (Conway et al., 2001). In further studies the proteins detected by MALDI-TOF MS were identified and resulted that most peaks in the spectral profiles are derived from ribosomal proteins (Ryzhov and Fenselau, 2001).

Bacterial identification can be carried out by either identifying ion biomarker masses that could be correlated with theoretically determined protein masses in databases or by comparing the whole spectral profile to a reference database (van Baar, 2000). In the first approach, bacterial strains are identified by determining the masses of biomarker proteins by MALDI-TOF MS and searching against a protein database by matching the masses against sequence-derived masses (Demirev et al., 2004). A number of studies have been carried out, determining protein biomarkers by MALDI-TOF MS for bacterial species identification (Demirev et al., 2004; Fagerquist et al., 2005). A critical challenge of protein database searches is the necessary high mass accuracy, since some proteins have very similar masses. Furthermore, identification is limited to well-characterized microorganisms with known protein sequences available in proteome databases (Dare, 2006).

The second approach, also named "fingerprint approach" is the most applied for bacterial identification. It relies on spectral differences of bacterial species and identification is carried out by comparison of the spectral profile of an unknown strain to a reference database of spectral profiles (Giebel et al., 2010; Mazzeo et al. 2006). This approach allows the differentiation of bacterial strains, due to the high specific spectral profiles, named "fingerprints", obtained. For this purpose, it is not necessary to identify the proteins but just to determine a number of characteristic peaks that are representative for the corresponding species and/or genus.

3. MALDI-TOF MS fingerprinting, a rapid and reliable method for bacterial identification in food

MALDI-TOF MS fingerprinting proved to be applicable for bacterial identification at the genus-, species- and even strain level. In routine bacterial identification in the clinical sector it demonstrated to be a rapid, cost-effective and accurate technique that achieved correct species identification of more than 92%. This is a significantly better result than conventional biochemical systems or even 16S rRNA sequencing for bacterial identification (Seng et al., 2009; Bizzini et al., 2010). Several authors agree that the costs of bacterial identification by MALDI-TOF MS fingerprinting is around two-thirds less than conventional methods, when taking into account the cost of materials and staff (Hsieh et al., 2008; Nassif, 2009). Furthermore, it has several advantages over other fast methods relying on genomics, such as DNA-microarrays, because fewer steps are necessary to achieve bacterial identification and thus, fewer errors are introduced along the analyzing process. Another advantage of MALDI-TOF mass fingerprinting is the effortless analysis of results, since no extensive data processing and statistical analysis is required, as it is the case in other rapid methods for bacterial identification, such as FT-IR and DNA-microarrays.

Until recently, MALDI-TOF MS techniques for the identification and typing of microorganisms remained confined to research laboratories and to certain species and the comparison with other species is limited by the accessibility of spectra. In the last years, the availability of MALDI-TOF MS devices and the reduction of costs, which enabled their use in either clinical, food or environmental microbiology laboratories, helped to increase the number of studies for the identification of different food pathogens. Thus, regarding to the application to routine bacterial identification, MALDI-TOF MS has shown to be a fast, reliable and cost-effective technique that has the potential to replace and/or complement conventional phenotypic identification for most bacterial strains isolated in clinical, food or environmental microbiology laboratories. Some authors conclude that the identification of microbial isolates by whole-cell mass spectrometry is one of the latest tools, forging a revolution in microbial diagnostics, with the potential of bringing to an end many of the time-consuming and man-power-intensive identification procedures that have been used for decades. However, to increase the reliability of the method, it should be taken into account that for routine identification of bacterial isolates, correct identification by MALDI-TOF MS at the species or strain level should be achieved (Bizzini et al., 2010).

In this sense, bacterial differentiation at the species level is not always possible with the commonly applied methods for bacterial identification. Thus, the analysis of the 16S rRNA gene resulted to be complicated in some cases due to the high similarity of sequences of species of the same genus, such as *Bacillus* spp. and *Pseudomonas* spp. However, the correct identification of the corresponding species is of great importance for food safety and quality, since the pathogenic and spoilage character can vary significantly. With MALDI-TOF MS fingerprinting a higher discrimination potential has been described, allowing the differentiation and correct identification of much close bacterial species and even strains of the same species (Keys et al., 2004; Donohue et al., 2006; Vargha et al., 2006).

In recent years, several reports have shown the feasibility of using MALDI-TOF MS for identifying microorganisms (Seng et al., 2009; Giebel et al., 2010). The detection and comparison of protein mass patterns has become a convenient tool for the rapid identification of bacteria, due to the high specific mass profiles obtained. It has to be mentioned that most studies of bacterial identification by MALDI-TOF MS fingerprinting are targeted at clinical diagnostics of bacterial strains associated with human infectious diseases.

In contrast, only few works have been done in the field of microbial food analysis by MALDI-TOF MS for the identification of foodborne pathogens and/or spoilers. These works included the classification and identification of several widespread pathogens causing food-borne diseases such as *Aeromonas A. hydrophila, Arcobacter spp., Campylobacter spp., Clostridium spp., Listeria spp., Salmonella spp., Staphylococcus spp., V. parahaemolyticus, Yersinia* spp., *Bacillus* spp. and species of the *Enterobacteriaceae* family (Bernardo et al., 2002; Bright et al., 2002; Keys et al., 2004; Mandrell et al., 2005; Donohue et al., 2006; Carbonnelle et al., 2007; Barbuddhe et al., 2008; Grosse-Herrenthey et al., 2008; Hazen et al., 2009; Alispahic et al., 2010; Ayyadurai et al., 2010; Dubois et al., 2010; Stephan et al., 2010; Stephan et al., 2011). Furthermore, some studies were aimed at the detection of foodborne pathogens and food spoilage bacteria, including genera such as *Escherichia, Yersinia, Proteus, Morganella, Salmonella, Staphylococcus, Micrococcus, Lactococcus, Pseudomonas, Leuconostoc* and *Listeria* (Mazzeo et al., 2006). In further studies an ample spectral library was created, including the main pathogenic and spoilage bacterial species potentially present in seafood (Böhme et al., 2010a; Fernández-No et al., 2010; Böhme et al., 2011b). These works included genera, such as *Acinetobacter, Aeromonas, Bacillus, Carnobacterium, Listeria, Pseudomonas, Shewanella, Staphylococcus, Stenotrophomonas, Vibrio* and genera of the *Enterobacteriaceae* family.

In Figure 1 spectral profiles of some important foodborne pathogens and spoilers are shown, demonstrating the high specificity. Böhme et al. (2010a, 2011b) also determined characteristic biomarker peaks for every studied genus and species. Such unique or characteristic peak masses can serve for the rapid identification of a bacterial genus and/or species. However, unequivocal identification can not be carried out based on a single biomarker protein, but under consideration of a number of characteristic mass patterns, representing the spectral fingerprint.

Furthermore, when working with microbial mixtures, such biomarkers become more important, since the presence or absence of unique peak patterns could lead to a conclusion of the present bacterial species. The detection of biomarker proteins by MALDI-TOF MS has been successfully applied for the identification of two bacterial species isolated from contaminated water, lettuce and cotton cloth (Holland et al., 2000). However, until now, the application of MALDI-TOF MS fingerprinting for microbial mixtures has not yet been demonstrated.

Another critical challenge of MALDI-TOF MS fingerprinting is that the classification of a bacterial genus or species, as well as the determination of unique biomarker patterns, is only possible in the frame of the content of the spectral reference library. However, the number of studies on bacterial species identification by MALDI-TOF MS in foodstuffs is continually increasing and so does the reliability of the identification.

Figure 2 shows the scheme of the whole process of bacterial identification by MALDI-TOF MS in food products. Bacteria are isolated from food samples and cultivated to obtain single colonies. Afterwards, the bacterial cells are lysed by an organic solvent and a strong acid, being the most applied ones Acetonitrile (ACN) and Trifluoracetic acid (TFA). Once obtained the spectral profiles for each bacterial strain, data analysis is carried out, including the extraction of representative peak mass lists and the comparison of spectral data, with the aim of bacterial discrimination. Furthermore, cluster analysis of the peak mass lists reveals phyloproteomic relationships between bacterial species, allowing the identification of unknown strains, as well as the typing of closely related strains. For the sample preparation and data analysis a number of different protocols and techniques have been described and are discussed in the following sections.

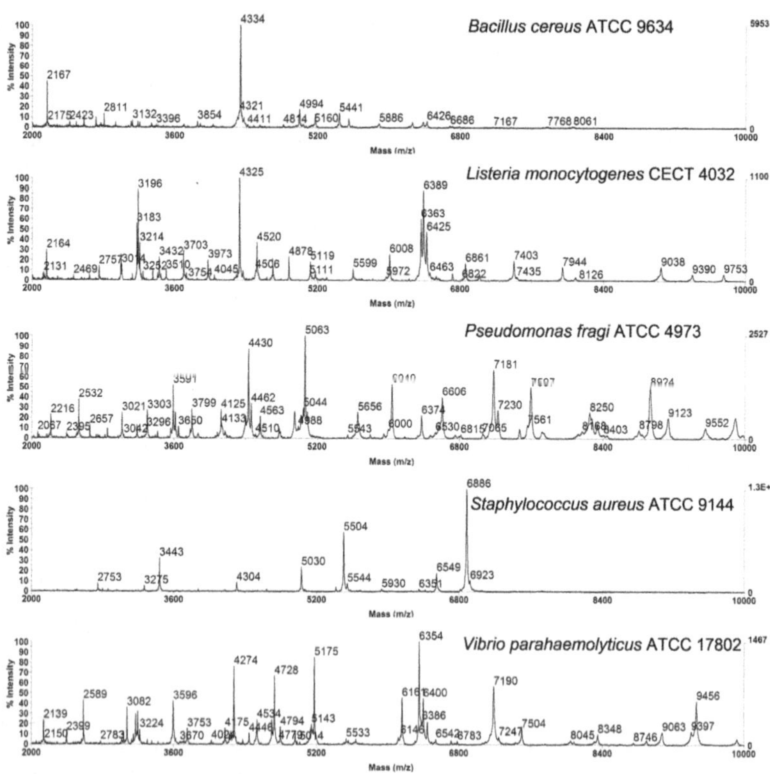

Fig. 1. MALDI-TOF MS profiles of some foodborne and spoilage bacteria.

For bacterial identification by MALDI-TOF MS fingerprinting, the spectral profile of the strain of interest is compared to a spectral library of reference strains. Several private databases have been created, including spectral profiles of more than 500 bacterial strains, such as The Spectral Archive And Microbial Identification System (Saramis™; *AnagnosTec* GmbH, Potsdam, Germany) (Erhard et al., 2008) and the Microbelynx bacterial identification system (*Waters* Corporation, Manchester, UK) (Keys et al., 2004; Dare, 2006). The MALDI Biotyper 2.0 (Bruker Daltonics) search against an ample database of more than 1800 bacterial species and new spectral profiles are being added on a daily basis. The database demonstrated to be applicable for the routine bacterial identification in the clinical sector, being a rapid, cost-effective and accurate technique that achieved correct species identification of 92% (Seng et al., 2009; Bizzini et al., 2010). As already mentioned, most studies, as well as these databases, are targeted at human pathogens causing infectious diseases. Nevertheless, the databases also include bacterial species that play an important role in food safety and quality and could represent important reference data for the identification of food pathogens and food spoilage bacterial species. However, the critical challenge of these databases is the limited availability. In this sense, it would be desirable a public database for the submission of the spectral information for each species that would allow the comparison with results from different researches and favoring a more precise method for identification of intact bacteria based on a huge amount of data.

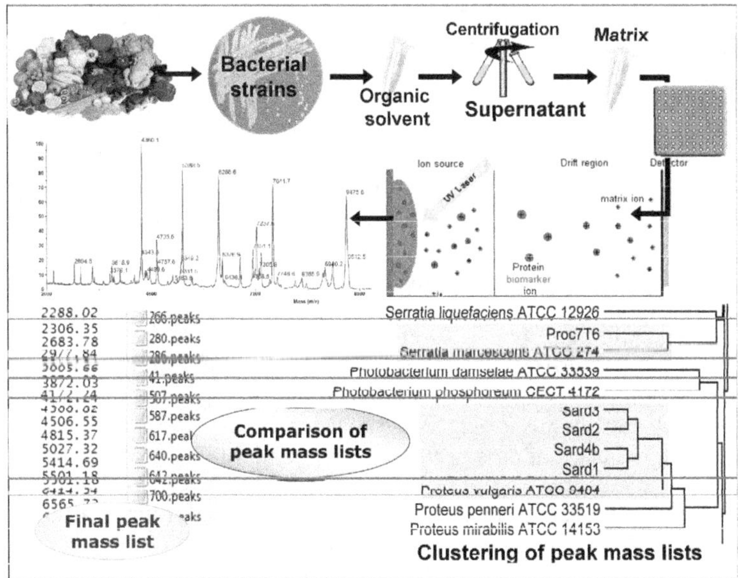

Fig. 2. Scheme of the protocol for the identification of foodborne and spoilage bacteria by MALDI-TOF MS fingerprinting.

Thus, a few attempts to start a public database have been achieved. Mazzeo et al. (2006) constructed a library containing spectra of 24 food-borne bacterial species, including *Escherichia spp., Yersinia spp., Proteus spp., Morganella spp., Salmonella spp., Staphylococcus spp., Micrococcus spp., Lactococcus spp., Pseudomonas spp., Leuconostoc spp.* and *Listeria spp.* Although, the spectral profiles and peak mass lists are freely available on the Web (http://bioinformatica.isa.cnr.it/Descr_Bact_Dbase.htm), the library only includes a few bacterial species important in food-borne diseases and/or food spoilage.

A reference library of mass spectral fingerprints of the main pathogenic and spoilage bacterial species, potentially present in seafood products has been created, including more than 50 bacterial species with interest in the food sector (Böhme et al., 2010a; Böhme et al., 2011b). In further studies, the library showed to be applicable for the correct identification of unknown bacterial strains isolated from commercial seafood products (Böhme et al., 2011a). It should be emphasized that the compiled reference library of seafood borne and spoilage bacterial species can be applied to any other foodstuff. The constituted spectral library may easily be enlarged by further bacterial species and strains that are of interest in the corresponding food product.

3.1 Sample preparation protocol

To allow the comparison of bacterial spectral fingerprints, the obtained spectra have to be representative and reproducible. For that, a standardized protocol has to be followed, beginning from sample preparation to instrumental parameters. It has been shown that spectral profiles were less sensitive to culture conditions but can show significant variability depending on the sample preparation protocol (Bernardo et al., 2002; Wunschel et al., 2005).

In the first studies of bacteria by MALDI-TOF MS, protein fractions were isolated from bacterial cells but shortly turned to the analysis of whole cells directly without any sample pre-treatment, called intact cell mass spectrometry (ICMS). Many authors focused on the optimization of the sample preparation protocol, with the aim to establish a standardized protocol to obtain specific and reproducible spectral profiles in a rapid and labor-saving way (Williams et al., 2003; Mazzeo et al., 2006; Liu et al., 2007). Nowadays, three different sample preparation protocols are commonly applied. Table 1 resumes these different methods, highlighting their advantages and disadvantages. The most rapid and labor-saving method is based on the direct application of bacterial biomass taken from culture plates to the MALDI-TOF MS sample plate. Afterwards, the bacterial cells are overlaid with the matrix solution (Bright et al., 2002). Apart from being the most rapid and labor-saving method, the direct spotting of biomass to the sample plate had several disadvantages. The difficulty in taking the correct amount of biomass complicates to get a homogenous distribution of the sample and matrix. Although, this technique was successfully applied for bacterial species identification (Keys et al., 2004; Erhard et al., 2008), it has been shown that spectra showed more noise and less peak resolution with this fast method, making difficult to obtain reproducible spectral profiles (Böhme et al., 2010b). Another sample preparation technique analyzed cell suspensions that were obtained after harvesting bacterial biomass in a solvent, including one or two washing steps and resuspension of the pellet in the matrix solution (Mazzeo et al., 2006; Vargha et al., 2006). The disadvantages of this method are the time-consuming washing steps and the loss of small soluble proteins. Some authors also described a similar sample preparation method, but where no washing step was applied and the bacterial colonies were harvested in a solvent to obtain cell suspensions (Carbonnelle et al., 2007).

Whole Cells	Whole Cell Suspension	Cell extract
Sample preparation		
Direct application of biomass to target well	Harvest biomass in organic solvent 1-2 cycles of Washing/ Centrifugation steps	Harvest biomass in organic solvent 1 Centrifugation step Analysis of supernatant
Advantages and disadvantages		
Very fast method	Time consuming washing steps	Very fast method
Less homogenous crystallization	Homogenous crystallization	Homogenous crystallization
Less reproducibility	Good reproducibility	Best reproducibility
More noise	More noise	Low noise
Less resolution of peaks	Good resolution of peaks	Best resolution of peaks

Table 1. Comparison of different sample preparation protocols.

In a third approach, based on the latter one, cell suspensions were centrifuged and spectra obtained by the analysis of the supernatant. When comparing to the spectra resulting from cell suspensions without centrifugation step, the spectral profiles obtained of the supernatant showed a better reproducibility, a higher resolution and less noise. The decreasing of noise, lessening the background, and the increase in resolution leads to more representative and characteristic peaks for each bacterial species improving the reproducibility (Böhme et al., 2010b). It should be mentioned that, in general terms, a more homogenous distribution of sample and matrix is expected with cell extracts, than with cell suspensions. In addition, this sample preparation method is rapid and effortless, since the extracts were obtained directly from cell cultures in just one dilution/centrifugation step. Although, when working with the extracts it should be expected to find small, soluble proteins, spectral profiles showed a high number of peaks, similar or even higher than those obtained by the analysis of whole cell suspensions. Nowadays, most applications of MALDI-TOF MS for bacterial identification analyzes bacterial cell extracts, obtained by just one dilution/ centrifugation step.

As mentioned before, for the direct comparison of spectral profiles, with the aim of bacterial differentiation, a strict standardized protocol has to be followed. It has been shown that, when applying the same culture conditions, sample preparation, matrix, organic solvents and MALDI-TOF MS analyzer, reproducible spectral profiles are obtained (Keys et al., 2004; Barbuddhe et al., 2008). Although, different protocols can lead to a high variability in the resulting spectral profiles, it has to be mentioned that some peaks were detected, even if different protocols were applied. Such characteristic and conserved peaks could serve as biomarker proteins for the corresponding genus and/or species.

3.2 Data analysis and phyloproteomics

For the identification of an unknown bacterial strain by MALDI-TOF MS fingerprinting, the spectral profile is compared to a spectral library of reference strains. The existing microbial identification databases MALDI Biotyper 2.0 (Bruker Daltonics), SaramisTM (AnagnosTec GmbH) and Microbelynx (Waters Corporation) include an amply library of spectral data. At the same time, the software for analyzing spectral data and carry out bacterial identification are incorporated to these databases. Identification of unknown microorganisms is performed by comparing their individual peak lists to the database. A matching score based on identified masses and their intensity correlation is generated and used for ranking of the results. Furthermore, based on the similarity scores dendrograms can be constructed and principle component analysis can be carried out. The critical challenges of these databases and software are that both are conditioned to the corresponding instrument and are not available for other investigators without paying high charges.

Thus, in a number of studies, smaller, "in house" libraries have been constructed. The difficulty of developing an "in house" database lays in the need for particular algorithms to analyze or compare obtained spectra and to carry out searches against the constructed reference library. Peak matching techniques eliminate the subjectivity of visual comparison. Jarman et al. (2003) developed an automated peak detection algorithm to extract representative mass ions from a fingerprint and to compare spectra to fingerprints in a reference library. This algorithm carries out the identification of an unknown spectrum by calculating a degree of matching and is robust to the variability in the ion intensity (Jarman et al., 2003). Other authors developed a software (BGP-database, available on http://sourceforge.net/projects/bgp) to analyze and compare spectral profiles, allowing the rapid identification. This software determines the best match between the tested strain and the reference strains of the database, taking into account a possible error of the *m/z* value (Carbonnelle et al., 2007).

In further studies, the freely available web-based application SPECLUST (http: //bioinfo.thep.lu.se/speclust.html) was used to extract representative peak masses and to obtain final peak mass lists for each bacterial strain. Later on, required mass lists can be compared and common peak masses defined. The web interface calculates the mass difference between two peaks taken from different peak lists and determines if the two peaks are identical after taking into account a certain measurement uncertainty (σ) and peak match score (s) (Alm et al., 2006). The web program resulted very fast, easy to handle, and could be extended by new spectral mass lists in a simple manner. Although, it was not possible to search an unknown spectrum directly against the library, comparison of peak mass lists could be carried out and common peaks determined with the aim of identifying a spectral profile of an unknown strain. The web-application was successfully applied to identify pathogenic and spoilage bacterial strains, isolated from commercial seafood products (Böhme et al., 2011a). In addition, the program allows the rapid determination of specific biomarker peaks and includes a clustering option.

Further bioinformatics programs, such as Statgraphics Plus 5.1 (Statpoint Technologies, inc., Warrenton, USA), offer a variety of functionalities. First, spectral data have to be transformed to a binary table, indicating the presence (1) and absence (0) of a peak mass. Afterwards, various algorithms for cluster analysis can be applied, as well as Principal Component Analysis (Böhme et al., 2011b). In a different study, the BioNumerics 6.0 software (Applied-Maths, Sint-Martens-Latem, Belgium) was used for data analysis and machine learning for bacterial identification by MALDI-TOF MS (De Bruyne et al., 2011).

Clustering of the spectral data obtained by MALDI-TOF MS represents another approach for bacterial identification and classification. Conway et al. (2001) introduced the term "Phyloproteomics" and the clustering of peak mass lists allowed a better visualization of similarities and differences of spectral comparison. The construction of a dendrogram based on mass spectral data is a rapid technique to analyze spectral profiles and to visualize spectral relations by grouping the obtained peak mass lists of bacterial strains. Thus, on one hand, clustering has been successfully applied for the differentiation and identification of bacterial strains at the genus and species level (Conway et al., 2001; Wunschel et al., 2005; Vargha et al., 2006; Carbonnelle et al., 2007; Böhme et al., 2010a; Dubois et al., 2010; Böhme et al., 2011b). On the other hand, clustering of mass spectral data has been applied as a typing method for the phyloproteomic study of different strains of the same species, with the aim to classify the strains (Siegrist et al., 2007; Teramoto et al., 2007). When comparing the dendrograms representing phyloproteomic relations to the phylogenetic trees, a high concordance were found by these authors. Since the peak patterns observed by MALDI-TOF MS are generally attributed to ribosomal proteins (Ryzhov and Fenselau, 2001), the similarity of the MALDI-TOF MS cluster to phylogenetic trees obtained by the analysis of ribosomal genes is not surprising (Dubois et al., 2010). However, in comparison to the sequence analysis of the 16S rRNA gene that is commonly used for phylogenetic studies, the classification of bacterial strains by MALDI-TOF MS fingerprinting resulted to be more discriminating. This is important for some genera, such as *Bacillus* and *Pseudomonas*, for that the differentiation at the species level is difficult with 16S rRNA analysis.

4. Conclusion

In order to ensure food safety and quality, the objective of this chapter is to review and update MALDI-TOF MS-based molecular methods that allow for the early detection and identification of the main pathogenic and spoilage bacteria. In recent years, molecular diagnosis has increased in importance, representing an attractive alternative to the traditional techniques of cultivation and characterization for the rapid detection and identification of microorganisms in food products that may produce foodborne illnesses.

MALDI-TOF MS fingerprinting proved to be a feasible, rapid and cost-effective technique for the classification and discrimination of bacterial strains. Besides the high potential in clinical diagnostics, it demonstrated to be applicable for the identification of unknown bacterial strains isolated from food samples. In addition, the methodology of MALDI-TOF MS fingerprinting can be applied to different foodstuff, since bacterial strains are isolated from the corresponding matrix before analysis and the constituted spectral library may easily be enlarged by adding bacterial species and strains that are of interest in the corresponding food product. The vast spectral data can be effectively examined by searching common peaks masses and cluster analysis. Characteristic peak patterns can be determined and serve as biomarker proteins for the rapid identification of an unknown spectrum. The creation of a dendrogram based on phyloproteomic relations, allow for the typing of closely related strains, extending conventional typing methods. When comparing phyloproteomic clustering with the classical taxonomy approach based on 16S rRNA gene analysis, high similarities were observed in the grouping of bacterial strains. However, genetic analysis does not always permit differentiation between species of the same genus due to the high similarity of DNA sequences, whereas spectral profiles obtained by MALDI-TOF analysis can reveal specificities for individual bacterial species. Thus, proteomic tools and phyloproteomic clustering resulted in more discrimination capability of different species.

The critical challenge of the MALDI-TOF MS fingerprinting approach for bacterial identification in food is the limited availability of reference data. A huge public database, such as the GenBank database of the NCBI, which facilitate the comparison of an unknown DNA sequence to millions of reference strains, is not available for mass spectral data obtained by MALDI-TOF MS. In future the main objective should be the publication of the created spectral libraries, including strains obtained from culture collections, as well as strains isolated from food samples, to allow other researchers the comparison of their spectra to the library and thus, a more accurate and rapid identification of bacteria in food.

5. Acknowledgements

This work was funded by project 10PXIB261045PR from Xunta de Galicia and by project AGL2010-19646 from the Spanish Ministry of Science and Technology. The work of K. Böhme and I.C. Fernández-No is supported by a "Maria Barbeito" and "Lucas Labrada" research contract from Xunta de Galicia.

6. References

Alispahic, M., Hummel, K., Jandreski-Cvetkovic, D., Nöbauer, K., Razzazi-Fazeli, E., Hess, M. & Hess, C. (2010). "Species-specific identification and differentiation of *Arcobacter, Helicobacter* and *Campylobacter* by full-spectral matrix-associated laser desorption/ionization time of flight mass spectrometry analysis." *Journal of Medical Microbiology* 59(3): 295-301.

Alm, R., Johansson, P., Hjerno, K., Emanuelsson, C., Ringnér, M. & Häkkinen, J. (2006). "Detection and identification of protein isoforms using cluster analysis of MALDI-MS mass spectra." *Journal of Proteome Research* 5(4): 785-792.

Ayyadurai, S., Flaudrops, C., Raoult, D. & Drancourt, M. (2010). "Rapid identification and typing of *Yersinia pestis* and other *Yersinia* species by matrix-assisted laser desorption/ionization time-of-flight (MALDI-TOF) mass spectrometry." *BMC Microbiology* 10(1): 285.

Barbuddhe, S.B., Maier, T., Schwarz, G., Kostrzewa, M., Hof, H., Domann, E., Chakraborty, T. & Hain, T. (2008). "Rapid identification and typing of *Listeria* species by matrix-assisted laser desorption ionization-time of flight mass spectrometry." *Applied and Environment Microbiology* 74(17): 5402-5407.

Bernardo, K., Pakulat, N., Macht, M., Krut, O., Seifert, H., Fleer, S., Hünger, F. & Krönke, M. (2002). "Identification and discrimination of *Staphylococcus aureus* strains using matrix-assisted laser desorption/ionization-time of flight mass spectrometry." *Proteomics* 2(6): 747-753.

Bizzini, A., Durussel, C., Bille, J., Greub, G. & Prod'hom, G. (2010). "Performance of matrix-assisted laser desorption ionization-time of flight mass spectrometry for identification of bacterial strains routinely isolated in a clinical microbiology laboratory." *Journal of Clinical Microbiology* 48(5): 1549-1554.

Blackburn, C.d.W. (2006). "*Food spoilage microorganisms.*" Woodhead Publishing.

Böhme, K., Fernández-No, I.C., Barros-Velázquez, J., Gallardo, J.M., Calo-Mata, P. & Cañas, B. (2010a). "Species differentiation of seafood spoilage and pathogenic Gram-negative bacteria by MALDI-TOF mass fingerprinting." *Journal of Proteome Research* 9(6): 3169-3183.

Böhme, K., Fernández-No, I.C., Barros-Velázquez, J., Gallardo, J.M., Cañas, B. & Calo-Mata, P. (2010b). "Comparative analysis of protein extraction methods for the identification of seafood-borne pathogenic and spoilage bacteria by MALDI-TOF mass spectrometry." *Analytical Methods* 2(12): 1941-1947.

Böhme, K., Fernández-No, I., Gallardo, J., Cañas, B. & Calo-Mata, P. (2011a). "Safety assessment of fresh and processed seafood products by MALDI-TOF mass fingerprinting." *Food and Bioprocess Technology* 4(6): 907-918.

Böhme, K., Fernández-No, I.C., Barros-Velázquez, J., Gallardo, J.M., Cañas, B. & Calo-Mata, P. (2011b). "Rapid species identification of seafood spoilage and pathogenic Gram-positive bacteria by MALDI-TOF mass fingerprinting." *Electrophoresis* 32(21): 2951-2965.

Bright, J.J., Claydon, M.A., Soufian, M. & Gordon, D.B. (2002). "Rapid typing of bacteria using matrix-assisted laser desorption ionisation time-of-flight mass spectrometry and pattern recognition software." *Journal of Microbiological Methods* 48(2-3): 127-138.

Carbonnelle, E., Beretti, J.-L., Cottyn, S., Quesne, G., Berche, P., Nassif, X. & Ferroni, A. (2007). "Rapid identification of *Staphylococci* isolated in clinical microbiology laboratories by matrix-assisted laser desorption ionization-time of flight mass spectrometry." *Journal of Clinical Microbiology* 45(7): 2156-2161.

Conway, G.C., Smole, S.C., Sarracino, D.A., Arbeit, R.D. & Leopold, P.E. (2001). "Phyloproteomics: species identification of *Enterobacteriaceae* using matrix-assisted laser desorption/ionization time-of-flight spectrometry." *Journal of Molecular Microbiology and Biotechnology* 3(1): 103-112.

Dare, D. (2006). Rapid bacterial characterization and identification by MALDI-TOF mass spectrometry. *Advanced Techniques in Diagnostic Microbiology*. Tang, Y.-W. & Stratton, C.W. New York, Springer Science+Business Media, LLC: 117-133.

De Bruyne, K., Slabbinck, B., Waegeman, W., Vauterin, P., De Baets, B. & Vandamme, P. (2011). "Bacterial species identification from MALDI-TOF mass spectra through data analysis and machine learning." *Systematic and Applied Microbiology* 34(1): 20-29.

Demirev, P.A., Feldman, A.B. & Lin, J.S. (2004). "Bioinformatics-based strategies for rapid microorganism identification by mass spectrometry." *John Hopkins APL Technical Digest* 25(1): 27-37.

Donohue, M.J., Smallwood, A.W., Pfaller, S., Rodgers, M. & Shoemaker, J.A. (2006). "The development of a matrix-assisted laser desorption/ionization mass spectrometry-based method for the protein fingerprinting and identification of *Aeromonas* species using whole cells." *Journal of Microbiological Methods* 65(3): 380-389.

Dubois, D., Leyssene, D., Chacornac, J.P., Kostrzewa, M., Schmit, P.O., Talon, R., Bonnet, R. & Delmas, J. (2010). "Identification of a variety of *Staphylococcus* species by matrix-assisted laser desorption ionization-time of flight mass spectrometry." *Journal of Clinical Microbiology* 48(3): 941-945.

Duerden, B.I., Towner, K.J. & Magee, J.T. (1998). Isolation, description and identification of bacteria. *Topley and Wilson's Microbiology and Microbial Infections: Systematic Bacteriology*. Collier, L.et al. London: 65-84.

Elizaquível, P., Chenoll, E. & Aznar, R. (2008). "A TaqMan-based real-time PCR assay for the specific detection and quantification of Leuconostoc mesenteroides in meat products." *FEMS Microbiology Letters* 278(1): 62-71.

Erhard, M., Hipler, U.-C., Burmester, A., Brakhage, A.A. & Wöstemeyer, J. (2008). "Identification of dermatophyte species causing onychomycosis and tinea pedis by MALDI-TOF mass spectrometry." *Experimental Dermatology* 17(4): 356-361.

Fagerquist, C.K., Miller, W.G., Harden, L.A., Bates, A.H., Vensel, W.H., Wang, G. & Mandrell, R.E. (2005). "Genomic and proteomic identification of a DNA-binding protein used in the "Fingerprinting" of *Campylobacter* species and strains by MALDI-TOF-MS protein biomarker analysis." *Analytical Chemistry* 77(15): 4897-4907.

FAO (2003). "Assuring food safety and quality : guidelines for strengthening national food control systems." *FAO food and nutrition paper, 0254-4725 ; 76*. Rome, Food and Agriculture Organization of the United Nations (FAO); World Health Organization (WHO).

Feng, P. (2007). Rapid methods for the detection of foodborne pathogens: Current and next-generation technologies. *Food Microbiology: Fundamentals and Frontiers*. Doyle, M.P. & Beuchat, L.R. Washington, D.C., ASM Press: 911-934.

Fernández-No, I.C., Böhme, K., Gallardo, J.M., Barros-Velázquez, J., Cañas, B. & Calo-Mata, P. (2010). "Differential characterization of biogenic amine-producing bacteria involved in food poisoning using MALDI-TOF mass fingerprinting." *Electrophoresis* 31(6): 1116-1127.

Giebel, R., Worden, C., Rust, S.M., Kleinheinz, G.T., Robbins, M., Sandrin, T.R., Allen, I.L., Sima, S. & Geoffrey, M.G. (2010). Microbial fingerprinting using matrix-assisted laser desorption ionization time-of-flight mass spectrometry (MALDI-TOF MS):

Applications and challenges. *Advances in Applied Microbiology*, Academic Press. 71: 149-184.

Gram, L., Ravn, L., Rasch, M., Bruhn, J.B., Christensen, A.B. & Givskov, M. (2002). "Food spoilage - interactions between food spoilage bacteria." *International Journal of Food Microbiology* 78(1-2): 79-97.

Grosse-Herrenthey, A., Maier, T., Gessler, F., Schaumann, R., Böhnel, H., Kostrzewa, M. & Krüger, M. (2008). "Challenging the problem of clostridial identification with matrix-assisted laser desorption and ionization-time-of-flight mass spectrometry (MALDI-TOF MS)." *Anaerobe* 14(4): 242-249.

Hazen, T.H., Martinez, R.J., Chen, Y., Lafon, P.C., Garrett, N.M., Parsons, M.B., Bopp, C.A., Sullards, M.C. & Sobecky, P.A. (2009). "Rapid identification of *Vibrio parahaemolyticus* by whole-cell matrix-assisted laser desorption ionization-time of flight mass spectrometry." *Applied Environmental Microbiology* 75(21): 6745-6756

Hein, I., Lehner, A., Rieck, P., Klein, K., Brandl, E. & Wagner, M. (2001). "Comparison of different approaches to quantify *Staphylococcus aureus* cells by Real-Time Quantitative PCR and application of this technique for examination of cheese." *Applied Environmental Microbiology* 67(7): 3122-3126.

Holland, R.D., Rafii, F., Heinze, T.M., Sutherland, J.B., Voorhees, K.J. & Lay Jr., J.O. (2000). "Matrix-assisted laser desorption/ionization time-of-flight mass spectrometric detection of bacterial biomarker proteins isolated from contaminated water, lettuce and cotton cloth." *Rapid Communications in Mass Spectrometry* 14(10): 911-917.

Hsieh, S.-Y., Tseng, C.-L., Lee, Y.-S., Kuo, A.-J., Sun, C.-F., Lin, Y.-H. & Chen, J.-K. (2008). "Highly efficient classification and identification of human pathogenic bacteria by MALDI-TOF MS." *Molecular and Cellular Proteomics* 7(2): 448-456.

Jarman, K.H., Daly, D.S., Anderson, K.K. & Wahl, K.L. (2003). "A new approach to automated peak detection." *Chemometrics and Intelligent Laboratory Systems* 69(1-2): 61-76.

Keys, C.J., Dare, D.J., Sutton, H., Wells, G., Lunt, M., McKenna, T., McDowall, M. & Shah, H.N. (2004). "Compilation of a MALDI-TOF mass spectral database for the rapid screening and characterisation of bacteria implicated in human infectious diseases." *Infection, Genetics and Evolution* 4(3): 221-242.

Klaenhammer, T.R., Pfeiler, E. & Duong, T. (2007). Genomics and proteomics of foodborne microorganisms. *Food Microbiology: Fundamentals and Frontiers*. Doyle, M.P. & Beuchat, L.R. Washington, D.C., ASM Press: 935-951.

Liu, H., Du, Z., Wang, J. & Yang, R. (2007). "Universal sample preparation method for characterization of bacteria by matrix-assisted laser desorption ionization-time of flight mass spectrometry." *Applied and Environmental Microbiology* 73(6): 1899-1907.

Malorny, B., Paccassoni, E., Fach, P., Bunge, C., Martin, A. & Helmuth, R. (2004). "Diagnostic Real-Time PCR for detection of *Salmonella* in food." *Applied Environmental Microbiology* 70(12): 7046-7052.

Mandrell, R.E., Harden, L.A., Bates, A., Miller, W.G., Haddon, W.F. & Fagerquist, C.K. (2005). "Speciation of *Campylobacter coli, C. jejuni, C. helveticus, C. lari, C. sputorum,* and *C. upsaliensis* by matrix-assisted laser desorption ionization-time of flight mass spectrometry." *Applied and Environmental Microbiology* 71(10): 6292-6307.

Mazzeo, M.F., Sorrentino, A., Gaita, M., Cacace, G., Di Stasio, M., Facchiano, A., Comi, G., Malorni, A. & Siciliano, R.A. (2006). "Matrix-assisted laser desorption ionization-time of flight mass spectrometry for the discrimination of food-borne microorganisms." *Applied and Environmental Microbiology* 72(2): 1180-1189.

Mohania, D., Nagpal, R., Kumar, M., Bhardwaj, A., Yadav, M., Jain, S., Marotta, F., Singh, V., Parkash, O. & Yadav, H. (2008). "Molecular approaches for identification and characterization of *lactic acid bacteria*." *Journal of Digestive Diseases* 9(4): 190-198.

Nassif, X. (2009). "Editorial commentary: A revolution in the identification of pathogens in clinical laboratories." *Clinical Infectious Diseases* 49(4): 552-553.

Ryzhov, V. & Fenselau, C. (2001). "Characterization of the protein subset desorbed by MALDI from whole bacterial cells." *Analytical Chemistry* 73(4): 746-750.

Sandt, C., Madoulet, C., Kohler, A., Allouch, P., De Champs, C., Manfait, M. & Sockalingum, G. (2006). "FT-IR microspectroscopy for early identification of some clinically relevant pathogens." *Journal of Applied Microbiology* 101 (4): 785-797.

Seng, P., Drancourt, M., Gouriet, F., La Scola, B., Fournier, P.-E., Rolain, Jean M. & Raoult, D. (2009). "Ongoing revolution in bacteriology: Routine identification of bacteria by matrix-assisted laser desorption ionization time-of-flight mass spectrometry." *Clinical Infectious Diseases* 49(4): 543-551.

Severgnini, M., Cremonesi, P., Consolandi, C., Bellis, G. & Castiglioni, B. (2011). "Advances in DNA microarray technology for the detection of foodborne pathogens." *Food and Bioprocess Technology* 4(6): 936-953.

Siegrist, T.J., Anderson, P.D., Huen, W.H., Kleinheinz, G.T., McDermott, C.M. & Sandrin, T.R. (2007). "Discrimination and characterization of environmental strains of *Escherichia coli* by matrix-assisted laser desorption/ionization time-of-flight mass spectrometry (MALDI-TOF-MS)." *Journal of Microbiological Methods* 68(3): 554-562.

Stephan, R., Cernela, N., Ziegler, D., Pflüger, V., Tonolla, M., Ravasi, D., Fredriksson-Ahomaa, M. & Hächler, H. (2011). "Rapid species specific identification and subtyping of *Yersinia enterocolitica* by MALDI-TOF Mass spectrometry." *Journal of Microbiological Methods* 87(2): 150-153.

Stephan, R., Ziegler, D., Pflüger, V., Vogel, G. & Lehner, A. (2010). "Rapid genus- and species-specific identification of *Cronobacter* spp. by matrix-assisted laser desorption ionization-time of flight mass spectrometry." *Journal of Clinical Microbiology* 48(8): 2846-2851.

Tauxe, R.V. (2002). "Emerging foodborne pathogens." *International Journal of Food Microbiology* 78(1-2): 31-41.

Teramoto, K., Sato, H., Sun, L., Torimura, M., Tao, H., Yoshikawa, H., Hotta, Y., Hosoda, A. & Tamura, H. (2007). "Phylogenetic classification of *Pseudomonas putida* strains by MALDI-MS using ribosomal subunit proteins as biomarkers." *Analytical Chemistry* 79(22): 8712-8719.

van Baar, B.L.M. (2000). "Characterisation of bacteria by matrix-assisted laser desorption/ionisation and electrospray mass spectrometry." *FEMS Microbiology Reviews* 24: 193-219.

Vargha, M., Takáts, Z., Konopka, A. & Nakatsu, C.H. (2006). "Optimization of MALDI-TOF MS for strain level differentiation of *Arthrobacter* isolates." *Journal of Microbiological Methods* 66(3): 399-409.

WHO. (2007). Fact Sheet Number 237: "Food safety and foodborne illness." World Health Organization (WHO), *www.who.int/mediacentre/factsheets/fs237/en/*.

Williams, T.L., Andrzejewski, D., Lay, J.O. & Musser, S.M. (2003). "Experimental factors affecting the quality and reproducibility of MALDI TOF mass spectra obtained from whole bacteria cells." *Journal of the American Society for Mass Spectrometry* 14(4): 342-351.

Wunschel, D.S., Hill, E.A., McLean, J.S., Jarman, K., Gorby, Y.A., Valentine, N. & Wahl, K. (2005). "Effects of varied pH, growth rate and temperature using controlled fermentation and batch culture on Matrix Assisted Laser Desorption/Ionization whole cell protein fingerprints." *Journal of Microbiological Methods* 62(3): 259-271.

3

Strategies for Iron Biofortification of Crop Plants

Mara Schuler and Petra Bauer
Dept. Biosciences-Plant Biology,
Saarland University,
Saarbrücken,
Germany

1. Introduction

Iron (Fe) is an essential element for all living organisms because of its property of being able to catalyze oxidation/reduction reactions. Fe serves as a prosthetic group in proteins to which it is associated either directly or through a heme or an iron-sulfur cluster. It exists in two redox states, the reduced ferrous Fe^{2+} and the oxidized Fe^{3+} form and is able to loose or gain an electron, respectively, within metalloproteins (e.g. Fe-S cluster or heme-Fe proteins). Such metalloproteins are involved in fundamental biochemical reactions like the electron transfer chains of respiration and photosynthesis, the biosynthesis of DNA, lipids and other metabolites, the detoxification of reactive oxygen species.

The cellular processes that involve Fe take place in distinct intracellular compartments like e.g. cytoplasm, mitochondria, plastids, cell walls, which therefore need to be provided with an adequate amount of Fe. Since this metal is involved in a wide range of essential processes, the undersupply with Fe leads to severe deficiency symptoms in the affected organism.

Fe deficiency is one of the most prevalent and most serious nutrient deficiencies threatening human health in the world, affecting approximately two billion people (de Benoist et al., 2008). Various physiological diseases, such as anaemia and some neurodegenerative diseases are triggered by Fe deficiency (Sheftela et al., 2011). Especially those countries are affected by Fe deficiency diseases, where people have low meat intake and the diets are mostly based on staple crops. Young children, pregnant and postpartum women are the most commonly and severely affected population groups, because of the high Fe demands of infant growth and pregnancy (de Benoist et al., 2008). Human health problems caused by Fe deficiency can be prevented by specific attention to food composition and by choosing a balanced diet with sufficient and bio-available Fe content.

Several possibilities exist to enrich the diet with bio-available Fe, which all have advantages and disadvantages. Supplementation of Fe in the diet is possible by supply of Fe chelates and salts in form of pills (Yakoob & Bhutta, 2011). However, formulations which are well tolerated by patients are expensive and particularly in underdeveloped areas of the world difficult to supply daily, as additional systems for purchasing, transport and distribution have to be established, associated with extra costs. The fortification of food products like flour with Fe salts is also effective (Best et al., 2011) and in place in some developed countries (Huma et al., 2007). Generally, an existing food industry is required for food processing, so that again supply is difficult in underdeveloped countries. The diversification of the diet with an emphasis on improvement of Fe-rich food crops like certain green leafy vegetables and legume seeds would be highly effective and desirable. In fact, it is actually the simplification of the diet with its low diversification that is the main cause of the micronutrient deficiency (Nair & Iyengar, 2009). The structure of agriculture, the green revolution and the need to supply sufficient food in light of a rapidly increasing world population had caused a concentration on calorie-rich carbohydrate-providing crops (Gopalan, 1996). Finally, the bio-fortification of staple crops is considered to be a very effective method which would reach many people even in underdeveloped countries (Bouis et al., 2011). A prerequisite is that the local staple crops are bio fortified so that they contain more and better available Fe. This can generally be reached by breeding, which is performed either by the breeding industry or by governmental agencies. The newly bred lines need to be distributed to and accepted by the local farmers. In any case, it seems that the prevention of Fe deficiency in the population of underdeveloped countries may strongly depend on governmental willingness, administration and regulation concerning the quality and quantity of food. It is clear that none of the above mentioned treatments is "cheap". Yet, the economic losses due to fatigue and neuronal dysfunctions might be far greater in negative value than the expected expenses to prevent these problems (Hunt, 2002). Therefore, the combat against Fe deficiency diseases is among the top priorities particularly listed by the WHO (de Benoist et al., 2008).

Here, we present some of the approaches for bio-fortification of crops with Fe. This report will focus on the underlying technological advances and our knowledge about the physiological processes leading to the enrichment of specific plant organs with Fe and their increased bio-availability.

2. Overview about Fe homeostasis in plants

The most important plants for nutrition of humans and mammals are the highly evolved flowering plants (angiosperms). These include the major crops and plant model organisms like rice, maize, legumes and *Arabidopsis thaliana*. Fe is found in all plant organs, which include roots, leaves, flowers, fruits with seeds, storage organs like tubers. Depending on the plant crop species and its use all these various parts can be edible, and in this case the concentrations of bio-available Fe should be high. Under natural conditions, all Fe of living organisms ultimately enters the nutrition chain via plant roots. In the soil, Fe mainly exists as Fe^{3+}, often bound as iron hydroxides in mineral soil particles (Marschner, 1995). Plants need a Fe concentration of 10^{-6} M for optimal growth, but the concentration of free Fe^{3+} in an aerobic, aqueous environment of the soil with a pH of 7 is about 10^{-17} M. At lower pH the solubility of Fe is increased, but a Fe^{3+} concentration of 10^{-6} M is reached at pH 3,3 (Hell & Stephan, 2003). 30% of the world`s crop land is too alkaline for optimal plant growth. Moreover, it appears that some staple crops, like rice, are especially susceptible to Fe deficiency. Under alkaline and calcareous soil conditions, bioavailable Fe concentrations are low in the soil despite of the abundance of this metal in the earth crust. To meet their demand for Fe, plants need to mobilize Fe in the soil by rendering it more soluble before they are able to take it up into their roots. Two effective Fe acquisition systems known as Strategy I and Strategy II have evolved in higher plants, based on reduction and chelation of Fe^{3+}, respectively (Römheld, 1987; Römheld & Marschner, 1986). The group of strategy I plants includes all dicotyledonous and all non-grass monocotyledonous plants. They acidify the soil, reduce Fe^{3+} and take up divalent Fe^{2+} via specific divalent metal transporters (Jeong & Guerinot, 2009; Morrissey & Guerinot, 2009). All monocotyledonous grasses are Strategy II plants, including all major cereal crop plants like rice (*Oryza sativa*), barley (*Hordum vulgare*), wheat (*Triticum aestivum*) and maize (*Zea mays*). These plants synthesize and secrete Fe^{3+}-chelating methionine derivatives termed phytosiderophores of the mugineic acid family and subsequently take up Fe^{3+}-phytosiderophore complexes (Jeong & Guerinot, 2009; Kobayashi et al., 2010; Morrissey & Guerinot, 2009). Fe reaches leaves mainly in complexed form with citrate through the xylem, which is a plant conductive tissue for water and mineral long-distance transport. Typical sink organs like immature organs receive Fe via the phloem pathway, which represents the conductive tissue for assimilates and signals. Inside plants, Fe is distributed to all tissues and cellular compartments through the activities of several different types of membrane-bound metal transport proteins (Curie et al., 2009; Jeong & Guerinot, 2009). Metal ions are predominant in a bound or chelated form inside cells to enhance solubility and transport but at the same time minimize toxicity effects of metal ions. In plants, oganic acids like citrate and malate, the amino acid histidine and the plant-specific methionine derivative nicotianamine are mainly involved in Fe transport and solubility (Briat et al., 2007; Callahan et al., 2006). Chelators for metals also include polypeptides such as phytochelatins (PCs) and metallothioneins (MTs) which are essentially involved in the tolerance to potentially toxic heavy metal ions (Hassinen et al., 2011; Pal & Rai, 2010). Fe can be stored in form of ferritin in the plastids which also serves to reduce oxidative stress (Briat et al., 2010b). In the vacuole Fe is frequently bound by phytic acid, which is composed of inositol esterified with phosphorous acid. The ionized form binds several mineral ions including Fe. It is present in cereal grains, nuts and leguminous seeds (Gibson et al., 2010).

In conclusion, plants contain a complex regulation network of genes which provide uptake, chelation, transport, sub-cellular distribution and the storage of Fe. Knowing these processes is the prerequisite for their manipulation in order to breed in the future high-quality nutritious crops.

3. Biofortification strategies

Bio-fortification designates the natural enrichment of plants with nutrients and health-promoting factors during their growth. Bio-fortification focuses on generating and breeding major staple food crops that would produce edible products enriched in bioavailable amounts of micronutrients, provitamin A carotenoids or several other known components that enhance nutrient use efficiency and are beneficial to human health (Hirschi, 2009).

The bio-fortification approach is interesting for staple crops that were mainly bred for carbohydrate content, processing characteristics and yield in the past decades, e.g. maize, wheat, rice and also some of the local plants like Cassava, potato and sweet potato. Elite lines highly performing in the field might on the other hand be poor in micronutrient contents (White & Broadley, 2009). Plants with a higher nutritional value can be produced by classical breeding. In this case, wild relatives or varieties with beneficial micronutrient content are selected and the respective trait crossed into the elite lines. This approach is labor-intensive, it can be aided by the usage of molecular markers that are closely linked with the traits of interest; in an optimal case, the molecular nature of the trait is known and can be followed directly with molecular PCR and sequencing technologies in the various breeding steps (Tester & Langridge, 2010; Welch & Graham, 2004). Alternatively, bio-fortified crops with new properties can be generated using gene technology in addition to classical breeding. In this case, the trait of interest is constructed *in vitro* using molecular cloning to combine promoters and genes that together confer the trait. These constructs are transferred into the crops, which could be achieved for example by biolistic methods based on the bombardment of plant cells with the DNA or using as tool *Agrobacterium tumefaciens.* The integration event of the DNA fragment conferring the new trait into the plant genome is selected, respective transgenic plants are generated and multiplied (Sayre et al., 2011; Shewry et al., 2008). Research on bio-fortification via classical breeding and/or gene technology-based breeding was stimulated by non-profit funding organizations, such as through the program HarvestPlus (http://www.harvestplus.org) (Bouis et al., 2011) and the Golden rice project (http://www.goldenrice.org) (Beyer, 2010). Bio-fortification thus became an agricultural and breeding tool to combat human malnutrition in the world.

For the Fe bio-fortification breeding, several challenges have to be overcome which can be mastered if scientists acquire a better understanding of the physiological mechanisms of plant metal homeostasis and political regulations allow for distributing such modified plants (Hotz & McClafferty, 2007). First, the plants have to increase Fe uptake. Depending on the soil properties, specific strategies for Fe mobilization in the soil have to be employed by the plants. Plants are then able to render Fe in the soil more soluble and bio-available to them. Second, Fe should accumulate in the edible parts of the plant such as seeds and fruits. These plant parts should act as effective sinks for Fe. Third, the nutrients should be preferentially stored in a form that renders them bioavailable for the human digestive system. Fe can be complexed with soluble organic ligands which would increase its bio-availability. However, some compounds like phytic acid can precipitate Fe and act as antinutrients if phytase is not provided.

First attempts to target physiological processes of Fe homeostasis have already been started to test the effect on bio-fortification. Moreover, assays are available to test for uptake of Fe from plant food items (Glahn et al., 2002; Lee et al., 2009; Maurer et al., 2010).

4. Examples for Fe biofortification research in plants

4.1 Reduction of phytic acid content

A successful approach for Fe bio-fortification relies on the reduction of Fe complex-forming metabolites that act as anti-nutrients, like tannins, a phenolic polymer, and phytic acid (Welch & Graham, 2004). Phytic acid (myo-inositol-1,2,3,4,5,6-hexakisphosphate; InsP6) comprises up to 80 % of the total seed phosphorus content and its dry mass may account for 1-2 % of seed weight (Hurrell, 2002). It accumulates as a phosphorous and mineral storage compound in globoids in the seeds of many staple crops, including legumes like soybean, cereal embryo and/or aleurone cells (Bohn et al., 2008). In developing countries, the prevalence of phytic acid in the plant-based diet is believed to contribute to the high rate of

Fe deficiency and anemia. On the other hand, reduction of phytic acid contents is also seen negative, since in a well-balanced diet it has health-promoting effects on the immune system and in preventing kidney stones (Shamsuddin, 2008). Phytic acid content can be reduced by disruption of its biosynthetic chain which would result in a "low phytic acid" (*lpa*) phenotype (Raboy, 2007; Rasmussen et al., 2010). Phytic acid is mainly synthesized from d-glucose-6-phosphate transformed first into 1d-*myo*-inositol-3-phosphate [Ins(3)P_1] (Loewus & Murthy, 2000). Several biochemical pathways seem to be involved in transforming Ins(3)P_1 to InsP_6 in plants, depending on the plant species (Bohn et al., 2008; Rasmussen et al., 2010). Furthermore, an ABC transporter is required for transport and compartmentalization in the final steps which can also be disrupted (Shi et al., 2007). Several mutant lines have been identified in various plant species including soybean (Hitz et al., 2002; Wilcox et al., 2000), maize (Pilu et al., 2003; Raboy et al., 2000), wheat (Guttieri *et al.*, 2004), rice (Larson et al., 2000; Liu et al., 2007) and Arabidopsis (Kim & Tai, 2011; Stevenson-Paulik et al., 2005). However, conventional breeding may result in strong phytic acid reduction and thereby in counteracting effects of such *lpa* mutants, like decreased germination and reduced seedling growth, if the effect takes place overall in the plants. Better mutants can be created using gene technology since only the late functions of the genes for phytate synthesis may be abolished and only in certain phases and organs during the life cycle of the plants by using specific promoters that allow expression of the transgenes under very controlled conditions (Kuwano et al., 2009; Kuwano et al., 2006).

Alternatively, the late stages of phytic acid biosynthesis and transport may be specifically targeted in mutants (Stevenson-Paulik et al., 2005). For example, two Arabidopsis genes for inositol polyphosphate kinases, ATIPK1 and ATIPK2, have been disrupted, which are required for the later steps of phytic acid synthesis. These mutants were found to produce 93 % less phytic acid in seeds, while seed yield and germination were not affected. It was however found that the loss of phytic acid precursors altered phosphate sensing.

An alternative approach may rely on the transformation of plants with phytase enzymes. Such enzymes are isolated from a multitude of different microorganisms, and heat-stability besides enzyme activity are important criteria to consider in the food processing procedure (Bohn et al., 2008; Rao et al., 2009).

Numerous examinations have to follow to find a solution to exclude negative influences of phytic acid as an anti-nutrient but sustain its positive effects on plant growth. It has to be investigated in future studies how useful phytate-reduced crops are for human Fe uptake.

4.2 Increase of ferritin content

Ferritins are multiprotein complexes consisting of ferritin peptide chains that are organized in globular manner to contain inside up to 4000 Fe^{3+} ions. Existing reports suggest that Fe is stored short- and long-term in ferritins and utilized for the accumulation of Fe-containing proteins. This way, ferritins supply Fe during developmental processes of plants, and some plant species contain high ferritin-Fe levels in seeds (Briat et al., 2010a). Ferritins also serve to alleviate oxidative stress (Briat et al., 2010b). However, not in every case high ferritin levels need to colocalize with high Fe levels in seeds (Cvitanich *et al.*, 2010). Ferritin-Fe is separated from other Fe-binding components by its protein coat and its localization inside plastids or mitochondria. Ferritins exist in all organisms as a store of Fe. Ferritins in general and ferritins in plant food items provide a high Fe bioavailability (Murray-Kolb et al., 2002; San Martin et al., 2008; Theil, 2004).

Ferritin genes were used in bio-fortification approaches. For example, leguminous ferritin genes, especially from soybean and bean, were over-expressed in plants, and subsequently an accumulation of ferritin protein was observed in the plants. Ferritins from legumes had been used since this plant family contains high ferritin levels in seeds, and the legume seeds serve in human and animal nutrition. Over-expression of ferritins in seeds and cereal grains resulted in an increased Fe content in these edible parts (Goto et al., 1999; Lucca et al., 2002). However, over-expression in vegetative tissues did not have this effect (Drakakaki et al., 2000), and in some cases even caused Fe deficiency symptoms (Van Wuytswinkel et al., 1999). Overall, ferritin over-expression has to studied in more detail and it may be needed to increase Fe uptake at the same time to have a full effect of Fe increases (Qu le et al., 2005).

Thus, research on the influence of ferritin on Fe accumulation and bio-availability as well as its effect on human Fe uptake revealed that this protein is a promising candidate for bio-fortification approaches if utilized in an appropriate manner in plants.

4.3 Increase of nicotianamine content

Nicotianamine is a key compound of metal homeostasis in plants. Nicotianamine is a non-proteinogenic amino acid derived from S-adenosyl methionine by the action of the enzyme nicotianamine synthase. Nicotianamine is able to bind a number of different metals including ferrous and ferric Fe, depending on the pH environment. Nicotianamine ensures that Fe remains soluble inside the cells. Thus, Fe can be transported to the multiple compartments, and Fe toxicity effects are reduced. Nicotianamine contributes to all important sub-processes of plant metal homeostasis: Mobilization and uptake, intercellular- and intracellular transport, sequestration, storage and detoxification of metals. Several studies presented positive effects of nicotianamine on Fe uptake and accumulation in seeds (Cheng et al., 2007; Douchkov et al., 2005; Douchkov et al., 2001; Klatte et al., 2009). Therefore, nicotianamine can be considered to be a potential bio-fortification factor for Fe in seeds and grains of crop plants. (Lee et al., 2009) showed that overexpression of a nicotianamine synthase gene, *OsNAS3*, resulted in an increase of Fe in leaves and seeds, and that in seeds a higher nicotianamine-Fe content was present. Moreover, it was found that these transgenic seeds provided a better source of dietary Fe than the wild type seeds (Lee et al., 2009). (Zheng et al., 2010) demonstrated by seed-specific expression of OsNAS1 that rice grains contained a higher amount of nicotianamine. These transgenic rice grains performed better in Fe utilization studies using human cells (Zheng et al., 2010). Other studies also indicated that simple overexpression of nicotianamine synthase genes may result in increased nicotianamine but not necessarily in augmented Fe utilization by the plants (Cassin et al., 2009). Excessive nicotianamine may restrict the availability of Fe when present in the apoplast (Cassin et al., 2009). It was also found that nicotianamine synthase overexpression can result in increased levels of Fe in leaves, but not consequently in seeds.

In conclusion, it can be stated that increased nicotianamine synthase gene expression can result in beneficial effects on bioavailability of Fe due to the chelator nicotianamine. However, care has to be taken on the site and amount of expression.

4.4 Combination of factors affecting bio-availability of Fe

The above studies suggested that targeting single genes may not necessarily result in an increased level of bio-available Fe. Combining multiple factors that affect bio-availability can be of further advantage. Such approaches have been tested. For example, rice grains expressing *Aspergillus* phytase, bean ferritin and a metallothionein were produced to contain higher levels of Fe in a form that might be bio-available (Lucca et al., 2002). In another study, maize plants were generated that expressed at the same time *Aspergillus* phytase and soybean ferritin in the endosperm of kernels (Drakakaki et al., 2005). These plants had an increased Fe content in seeds by 20-70% and nearly no phytate. Very interestingly, such kernels proved advantageous in bio-availability studies to human cells (Drakakaki et al., 2005).

(Wirth et al., 2009) produced rice plants simultaneously expressing three transgenes, namely a bean ferritin gene, Arabidopsis nicotianamine synthase gene *AtNAS1* and a phytase. Combined ferritin and nicotianamine over-production resulted in a stronger increase of Fe content in the endosperm of grains than was achieved in transgenic approaches with single genes (Wirth et al., 2009).

Thus, attempts to increase bioavailable Fe in seeds are becoming more successful, and combining multiple targets for breeding of Fe efficiency and Fe bio-availability seems to be the key.

4.5 Breeding for novel traits

The above presented approaches rely on the targeting of known components of plant Fe homeostasis mainly in gene technological approaches. An alternative non-transgenic approach is to use the genetic pool of germplasm collections to screen for plant lines that are Fe-efficient and have a high bio-availability of Fe. Such genetic traits can be mapped and backcrossed into the local elite varieties. An advantage of this genetic screening method is that no assumption about the physiology of the traits needs to be made beforehand. Due to the power of modern DNA sequencing the new genes and alleles of interest can eventually be molecularly identified, such as in the case of a transcription factor gene affecting seed micronutrient content (Uauy et al., 2006). In these cases, the power of natural genetic variation is utilized which is based on the natural selection of the best available traits that evolved in the germplasm collection, frequently based on the interplay of multiple genes and specific alleles (quantitative traits). As an example, plant breeders have begun screening for mineral content variation in collections of for example wild wheat (Chatzav *et al.*, 2010), rice (Gregorio et al., 2000) and bean (Blair et al., 2010). Furthermore, recombinant inbred lines generated from the original cross of two distantly related inbred lines may help in identifying and mapping of single and quantitative trait loci, for example in wheat (Peleg et al., 2009) and Medicago (Sankaran et al., 2009). In a different approach, cellular Fe uptake and bio-availability analyses have been used to screen rice or maize lines with novel traits not previously associated with known components of Fe usage (Glahn et al., 2002; Lung'aho et al., 2011).

5. Conclusion

Bio-fortification of crops with micronutrients contributes to the improvement of food quality and may help reducing the prevalent disease of Fe deficiency anemia world-wide. Multiple approaches using cereals and other crops have been tested and proven successful. It will remain as a challenge in the future to further improve details of these procedures, e.g. by exchanging isoforms of the genes, alleles, and new promoters in the case of transgenic approaches. Genetic breeding approaches can be improved by selecting novel recombinant inbred lines and new germplasm for testing. In some studies, the newly generated plant lines have not only been analyzed at plant physiological level for increased Fe content and gene/transgene activity but also for their capacity to augment Fe bio-availability to human epithelial cells (Drakakaki et al., 2005; Zheng et al., 2010) or to cure Fe deficiency anemia (Lee et al., 2009). Such bio-availability studies need to be performed routinely and also used in screening procedures to provide criteria for selection of the best plant lines.

6. References

Abadía, J., Vazquez, S., Rellan-Alvarez, R., El-Jendoubi, H., Abadia, A., Alvarez-Fernandez, A. and Lopez-Millan, A.F. (2011). Towards a knowledge-based correction of iron chlorosis. *Plant Physiol. Biochem.*, Vol. 49, pp. 482-471

Best, C., Neufingerl, N., Del Rosso, J.M., Transler, C., van den Briel, T. and Osendarp, S. (2011). Can multi-micronutrient food fortification improve the micronutrient status, growth, health, and cognition of schoolchildren? A systematic review. *Nutr. Rev.*, Vol. 69, pp. 186-204

Beyer, P. (2010). Golden Rice and 'Golden' crops for human nutrition. *Nat. Biotechnol.*, Vol. 27, pp. 478-481

Blair, M.W., Knewtson, S.J., Astudillo, C., Li, C.M., Fernandez, A.C. and Grusak, M.A. (2010). Variation and inheritance of iron reductase activity in the roots of common bean (*Phaseolus vulgaris* L.) and association with seed iron accumulation QTL. *BMC Plant Biol.*, Vol. 10, pp. 215

Bohn, L., Meyer, A.S. and Rasmussen, S.K. (2008). Phytate: impact on environment and human nutrition. A challenge for molecular breeding. *J. Zhejiang Univ. Sci. B*, Vol. 9, pp. 165-191

Bouis, H.E., Hotz, C., McClafferty, B., Meenakshi, J.V. and Pfeiffer, W.H. (2011). Biofortification: a new tool to reduce micronutrient malnutrition. *Food Nutr. Bull.* , Vol. 32, pp. S31-40

Briat, J.F., Curie, C. and Gaymard, F. (2007). Iron utilization and metabolism in plants. *Curr. Opin. Plant Biol.*, Vol. 10, pp. 276-282

Briat, J.F., Duc, C., Ravet, K. and Gaymard, F. (2010a). Ferritins and iron storage in plants. *Biochim. Biophys. Acta.*, Vol. 1800, pp. 806-814

Briat, J.F., Ravet, K., Arnaud, N., Duc, C., Boucherez, J., Touraine, B., Cellier, F. and Gaymard, F. (2010b). New insights into ferritin synthesis and function highlight a link between iron homeostasis and oxidative stress in plants. *Ann. Bot.*, Vol. 105, pp. 811-822

Callahan, D.L., Baker, A.J.M., Kolev, S.D. and Wedd, A.G. (2006). Metal ion ligands in hyperaccumulating plants. *Journal of Biological Inorganic Chemistry*, Vol. 11, pp. 2-12

Cassin, G., Mari, S., Curie, C., Briat, J.F. and Czernic, P. (2009). Increased sensitivity to iron deficiency in *Arabidopsis thaliana* overaccumulating nicotianamine. *J. Exp. Bot.*, Vol. 60, pp. 1249-1259

Chatzav, M., Peleg, Z., Ozturk, L., Yazici, A., Fahima, T., Cakmak, I. and Saranga, Y. (2010). Genetic diversity for grain nutrients in wild emmer wheat: potential for wheat improvement. *Ann. Bot.*, Vol. 105, pp. 1211-1220

Cheng, L.J., Wang, F., Shou, H.X., Huang, F.L., Zheng, L.Q., He, F., Li, J.H., Zhao, F.J., Ueno, D., Ma, J.F. and Wu, P. (2007). Mutation in nicotianamine aminotransferase stimulated the Fe(II) acquisition system and led to iron accumulation in rice. *Plant Physiol.*, Vol. 145, pp. 1647-1657

Curie, C., Cassin, G., Couch, D., Divol, F., Higuchi, K., Le Jean, M., Misson, J., Schikora, A., Czernic, P. and Mari, S. (2009). Metal movement within the plant: contribution of nicotianamine and yellow stripe 1-like transporters. *Ann. Bot.*, Vol. 103, pp. 1-11

Cvitanich, C., Przybyłowicz, W.J., Urbanski, D.F., Jurkiewicz, A.M., Mesjasz-Przybyłowicz, J., Blair, M.W., Astudillo, C., Jensen, E.Ø. and Stougaard, J. (2010). Iron and ferritin accumulate in separate cellular locations in Phaseolus seeds. *BMC Plant Biol.*, Vol. 10, pp. 26

de Benoist, B., McLean, E., Egli, I. and Cogswell, M. (2008). Worldwide prevalence of anaemia 1993-2005. *ISBN: 978 92 4 159665 7,*

Douchkov, D., Gryczka, C., Stephan, U.W., Hell, R. and Baumlein, H. (2005). Ectopic expression of nicotianamine synthase genes results in improved iron accumulation and increased nickel tolerance in transgenic tobacco. *Plant Cell Envir.*, Vol. 28, pp. 365-374

Douchkov, D., Hell, R., Stephan, U.W. and Baumlein, H. (2001). Increased iron efficiency in transgenic plants due to ectopic expression of nicotianamine synthase. *Plant Nutr.*, Vol. 92, pp. 54-55

Drakakaki, G., Christou, P. and Stöger, E. (2000). Constitutive expression of soybean ferritin cDNA in transgenic wheat and rice results in increased iron levels in vegetative tissues but not in seeds. *Transgenic Res.*, Vol. 9, pp. 445-452

Drakakaki, G., Marcel, S., Glahn, R.P., Lund, E.K., Pariagh, S., Fischer, R., Christou, P. and Stoger, E. (2005). Endosperm-specific co-expression of recombinant soybean ferritin and *Aspergillus phytase* in maize results in significant increases in the levels of bioavailable iron. *Plant Mol. Biol.*, Vol. 59, pp. 869-880

Gibson, R.S., Bailey, K.B., Gibbs, M. and Ferguson, E.L. (2010). A review of phytate, iron, zinc, and calcium concentrations in plant-based complementary foods used in low-income countries and implications for bioavailability. *Food Nutr. Bull.*, Vol. 31, pp. S134-146

Glahn, R.P., Cheng, Z., Welch, R.M. and Gregorio, G.B. (2002). Comparison of iron bioavailability from 15 rice genotypes: studies using an in vitro digestion/caco-2 cell culture model. *J. Agric. Food Chem.*, Vol. 50, pp. 3586-3591

Gopalan, C. (1996). Current food and nutrition situation in south Asian and south-east Asian countries. *Biomed. Environ. Sci.*, Vol. 9, pp. 102-116

Goto, F., Yoshihara, T., Shigemoto, N., Toki, S. and Takaiwa, F. (1999). Iron fortification of rice seed by the soybean ferritin gene. *Nat. Biotechnol.*, Vol. 17, pp. 282-286

Gregorio, G.B., Senadhira, D., Htut, T. and Graham, R.D. (2000). Breeding for trace mineral density in rice. *Food Nutr. Bull.*, Vol. 21, pp. 382-386

Guttieri, M., Bowen, D., Dorsch, J.A., Raboy, V. and Souza, E. (2004). Identification and characterization of a low phytic acid wheat. *Crop Sci.*, Vol. 44, pp. 418–424

Hassinen, V.H., Tervahauta, A.I., Schat, H. and Kärenlampi, S.O. (2011). Plant metallothioneins--metal chelators with ROS scavenging activity? *Plant Biol.*, Vol. 13, pp. 225-232

Hell, R. and Stephan, U.W. (2003). Iron uptake, trafficking and homeostasis in plants. *Planta,* Vol. 216, pp. 541-551

Hirschi, K.D. (2009). Nutrient biofortification of food crops. *Annu. Rev. Nutr.*, Vol. 29, pp. 401-421

Hitz, W.D., Carlson, T.J., Kerr, P.S. and Sebastian, S.A. (2002). Biochemical and molecular characterization of a mutation that confers a decreased raffinosaccharide and phytic acid phenotype on soybean seeds. *Plant Physiol.*, Vol. 128, pp. 650-660

Hotz, C. and McClafferty, B. (2007). From harvest to health: challenges for developing biofortified staple foods and determining their impact on micronutrient status. *Food Nutr. Bull.*, Vol. 28, pp. S271-279

Huma, N., Salim-Ur-Rehman, Anjum, F.M., Murtaza, M.A. and Sheikh, M.A. (2007). Food fortification strategy--preventing iron deficiency anemia: a review. *Crit. Rev. Food Sci. Nutr.* , Vol. 47, pp. 259-265

Hunt, J.M. (2002). Reversing productivity losses from iron deficiency: the economic case. *J Nutr.*, Vol. 132, pp. 794S 801S

Hurrell, R.F. (2002). Fortification: Overcoming Technical and Practical Barriers. *J. Nutr.*, Vol. 132, pp. 806S--812

Jeong, J. and Guerinot, M.L. (2009). Homing in on iron homeostasis in plants. *Trends Plant Sci.*, Vol. 14, pp. 280-285

Kim, S.I. and Tai, T.H. (2011). Identification of genes necessary for wild-type levels of seed phytic acid in Arabidopsis thaliana using a reverse genetics approach. *Mol. Genet. Genom.*, Vol. 286, pp. 119-133

Klatte, M., Schuler, M., Wirtz, M., Fink-Straube, C., Hell, R. and Bauer, P. (2009). The analysis of Arabidopsis nicotianamine synthase mutants reveals functions for nicotianamine in seed iron loading and iron deficiency responses. *Plant Physiol.*, Vol. 150, pp. 257-271

Kobayashi, T., Nakanishi, H. and Nishizawa, N.K. (2010). Recent insights into iron homeostasis and their application in graminaceous crops. *Proc. Japan. Acad. Ser. B Phys. Biol .Sci.*, Vol. 86, pp. 900-913

Kuwano, M., Mimura, T., Takaiwa, F. and Yoshida, K.T. (2009). Generation of stable 'low phytic acid' transgenic rice through antisense repression of the 1D-myo-inositol 3-phosphate synthase gene (RINO1) using the 18-kDa oleosin promoter. *Generation of stable 'low phytic acid' transgenic rice through antisense repression of the 1D-myo-inositol 3-phosphate synthase gene (RINO1) using the 18-kDa oleosin promoter*, Vol. 7, pp. 96-105

Kuwano, M., Ohyama, A., Tanaka, Y., Mimura, T., Takaiwa, F. and Yoshida, K.T. (2006). Molecular breeding for transgenic rice with low-phytic-acid phenotype through manipulating myo-inositol 3-phosphate synthase gene. *Mol. Breed.*, Vol. 18, pp. 263-272

Larson, S.R., Rutger, J.N., Young, K.A. and Raboy, V. (2000). Isolation and genetic mapping of a non-lethal rice (Oryza sativa L.) low phytic acid 1 mutation. *Isolation and genetic mapping of a non-lethal rice (Oryza sativa L.) low phytic acid 1 mutation*, Vol. 40, pp. 1397–1405

Lee, S., Jeon, U.S., Lee, S.J., Kim, Y.-K., Persson, D.P., Husted, S., Schjorring, J.K., Kakei, Y., Masuda, H., Nishizawa, N.K. and An, G. (2009). Iron fortification of rice seeds through activation of the nicotianamine synthase gene. *Proc. Natl. Acad. Sci. USA*, Vol. 106, pp. 22014--22019

Liu, Q.L., Xu, X.H., Ren, X.L., Fu, H.W., Wu, D.X. and Shu, Q.Y. (2007). Generation and characterization of low phytic acid germplasm in rice (*Oryza sativa* L.). *Theor. Appl. Genet.*, Vol. 114, pp. 803–814

Loewus, F.A. and Murthy, P.P.N. (2000). myo-Inositol metabolism in plants. *Plant Sci.*, Vol. 150, pp. 1-19

Lucca, P., Hurrell, R. and Potrykus, I. (2002). Fighting iron deficiency anemia with iron-rich rice. *J. Am. Coll. Nutr.*, Vol. 21, pp. 184S-190S

Lung'aho, M.G., Mwaniki, A.M., Szalma, S.J., Hart, J.J., Rutzke, M.A., Kochian, L.V., Glahn, R.P. and Hoekenga, O.A. (2011). Genetic and physiological analysis of iron biofortification in maize kernels. *PLoS One*, Vol. 6, pp. e20429

Marschner, H. (1995). Mineral Nutrition of Plants. *Academic Press, Boston,*

Maurer, F., Daum, N., Schaefer, U.F., Lehr, C.M. and Bauer, P. (2010). Plant genetic factors for iron homeostasis affect iron bioavailability in Caco-2 cells. *Food Res. Intl.*, Vol. 43, pp. 1661-1665

Morrissey, J. and Guerinot, M.L. (2009). Iron uptake and transport in plants: the good, the bad, and the ionome. *Chem. Rev.*, Vol. 109, pp. 4553-4567

Murray-Kolb, L.E., Takaiwa, F., Goto, F., Yoshihara, T., Theil, E.C. and Beard, J.L. (2002). Transgenic rice is a source of iron for iron-depleted rats. *J. Nutr.*, Vol. 132, pp. 957-960

Nair, K.M. and Iyengar, V. (2009). Iron content, bioavailability & factors affecting iron status of Indians. *Indian J. Med. Res.*, Vol. 130, pp. 634-645

Pal, R. and Rai, J.P. (2010). Phytochelatins: peptides involved in heavy metal detoxification. *Appl. Biochem. Biotechnol.*, Vol. 160, pp. 945-963

Peleg, Z., Cakmak, I., Ozturk, L., Yazici, A., Jun, Y., Budak, H., Korol, A.B., Fahima, T. and Saranga, Y. (2009). Quantitative trait loci conferring grain mineral nutrient concentrations in durum wheat x wild emmer wheat RIL population. *Theor. Appl. Genet.*, Vol. 119, pp. 353-369

Pilu, R., Panzeri, D., Gavazzi, G., Rasmussen, S.K., Consonni, G. and Nielsen, E. (2003). Phenotypic, genetic and molecular characterization of a maize low phytic acid mutant (*Lpa 241*). *Theor. Appl. Genet.*, Vol. 107, pp. 980–987

Qu le, Q., Yoshihara, T., Ooyama, A., Goto, F. and Takaiwa, F. (2005). Iron accumulation does not parallel the high expression level of ferritin in transgenic rice seeds. *Planta*, Vol. 222, pp. 225-233

Raboy, V. (2007). The ABCs of low-phytate crops. *Nat. Biotechnol.*, Vol. 25, pp. 874-875

Raboy, V., Gerbasi, P.F., Young, K.A., Stoneberg, S.D., Pickett, S.G., Bauman, A.T., Murthy, P.P., Sheridan, W.F. and Ertl, D.S. (2000). Origin and seed phenotype of maize *low phytic acid 1-1* and *low phytic acid 2-1. Plant Physiol.*, Vol. 124, pp. 355-368

Rao, D.E., Rao, K.V., Reddy, T.P. and Reddy, V.D. (2009). Molecular characterization, physicochemical properties, known and potential applications of phytases: An overview. *Crit. Rev. Biotechnol.*, Vol. 29, pp. 182-198

Rasmussen, S.K., Ingvardsen, C.R. and Torp, A.M. (2010). Mutations in genes controlling the biosynthesis and accumulation of inositol phosphates in seeds. *Biochem. Soc. Trans.*, Vol. 38, pp. 689-694

Römheld, V. (1987). Different strategies for iron acquisition in higher plants. *Physiol. Plant.*, Vol. 70, pp. 231-234

Römheld, V. and Marschner, H. (1986). Different strategies in higher plants in mobilization and uptake of iron. *J. Plant. Nutr.*, Vol. 9, pp. 695-713

San Martin, C.D., Garri, C., Pizarro, F., Walter, T., Theil, E.C. and Núñez, M.T. (2008). Caco-2 intestinal epithelial cells absorb soybean ferritin by mu2 (AP2)-dependent endocytosis. *J. Nutr.*, Vol. 138, pp. 659-666

Sankaran, R.P., Huguet, T. and Grusak, M.A. (2009). Identification of QTL affecting seed mineral concentrations and content in the model legume *Medicago truncatula*. *Theor. Appl. Genet.* , Vol. 119, pp. 241-253

Sayre, R., Beeching, J.R., Cahoon, E.B., Egesi, C., Fauquet, C., Fellman, J., Fregene, M., Gruissem, W., Mallowa, S., Manary, M., Maziya-Dixon, B., Mbanaso, A., Schachtman, D.P., Siritunga, D., Taylor, N., Vanderschuren, H. and Zhang, P. (2011). The BioCassava plus program: biofortification of cassava for sub-Saharan Africa. *Annu. Rev. Plant Biol.*, Vol. 62, pp. 251-272

Shamsuddin, A.M. (2008). Demonizing phytate. *Nat. Biotechnol.*, Vol. 26, pp. 496-497

Sheftela, A.D., Mason, A.B. and Ponka, P. (2011). The long history of iron in the Universe and in health and disease. *Biochim. Biophys. Acta*, pp. doi.org/10.1016/j.bbagen.2011.1008.1002

Shewry, P.R., Jones, H.D. and Halford, N.C. (2008). Plant biotechnology: transgenic crops. *Adv. Biochem. Eng. Biotechnol.*, Vol. 111, pp. 149-186

Shi, J., Wang, H., Schellin, K., Li, B., Faller, M., Stoop, J.M., Meeley, R.B., Ertl, D.S., Ranch, J.P. and Glassman, K. (2007). Embryo-specific silencing of a transporter reduces phytic acid content of maize and soybean seeds. *Nat. Biotechnol.*, Vol. 25, pp. 930-937

Stevenson-Paulik, J., Bastidas, R.J., Chiou, S.T., Frye, R.A. and York, J.D. (2005). Generation of phytate-free seeds in Arabidopsis through disruption of inositol polyphosphate kinases. *Proc. Natl. Acad. Sci. USA*, Vol. 102, pp. 12612--12617

Tester, M. and Langridge, P. (2010). Breeding technologies to increase crop production in a changing world. *Science*, Vol. 327, pp. 818-822

Theil, E.C. (2004). Iron, ferritin, and nutrition. *Annual review of nutrition*, Vol. 24, pp. 327--343

Uauy, C., Distelfeld, A., Fahima, T., Blechl, A. and Dubcovsky, J. (2006). A *NAC* gene regulating senescence improves grain protein, zinc, and iron content in wheat. *Science*, Vol. 314, pp. 1298-1301

Van Wuytswinkel, O., Vansuyt, G., Grignon, N., Fourcroy, P. and Briat, J.F. (1999). Iron homeostasis alteration in transgenic tobacco overexpressing ferritin. *Plant J.*, Vol. 17, pp. 93-97

Welch, R.M. and Graham, R.D. (2004). Breeding for micronutrients in staple food crops from a human nutrition perspective. *J. Exp. Bot.*, Vol. 55, pp. 353-364

White, P.J. and Broadley, M.R. (2009). Biofortification of crops with seven mineral elements often lacking in human diets--iron, zinc, copper, calcium, magnesium, selenium and iodine. *New Phytol.*, Vol. 182, pp. 49-84

Wilcox, J.R., Premachandra, G.S., Young, K.A. and Raboy, V. (2000). Isolation of high inorganic P, low-phytate soybean mutants. *Crop Sci.*, Vol. 40, pp. 1601–1605

Wirth, J., Poletti, S., Aeschlimann, B., Yakandawala, N., Drosse, B., Osorio, S., Tohge, T., Fernie, A.R., Günther, D., Gruissem, W. and Sautter, C. (2009). Rice endosperm iron biofortification by targeted and synergistic action of nicotianamine synthase and ferritin. *Plant Biotechnol. J.* , Vol. 7, pp. 631-644

Yakoob, M.Y. and Bhutta, Z.A. (2011). Effect of routine iron supplementation with or without folic acid on anemia during pregnancy. *BMC Publ. Health*, Vol. 11, pp. Suppl 3:S21

Zheng, L., Cheng, Z., Ai, C., Jiang, X., Bei, X., Zheng, Y., Glahn, R.P., Welch, R.M., Miller, D.D., Lei, X.G. and Shou, H. (2010). Nicotianamine, a novel enhancer of rice iron bioavailability to humans. *PloS One*, Vol. 5, pp. e10190

4

Raman Spectroscopy: A Non-Destructive and On-Site Tool for Control of Food Quality?

S. Hassing[1], K.D. Jernshøj[2] and L.S. Christensen[3]
[1]*Faculty of Engineering, Institute of Technology and Innovation, University of Southern Denmark,*
[2]*Faculty of Science, Department of Biochemistry and Molecular Biology, Celcom, University of Southern Denmark,*
[3]*Kaleido Technology, Denmark*

1. Introduction

In recent years there has been an increasing focus from the consumers on food quality i.e. unwanted substances such as bacteria, pesticides, drug residues and additives as well as on food composition including nutritional value, healthy additives, antioxidants and the contents of selected fatty acids. This is also reflected in an increasing interest for organic food products. It therefore seems appropriate to develop substance specific, non-destructive and fast measuring techniques that can be used close to the consumer, for monitoring different properties of food products.

Raman spectroscopy is an example of a fast, non-destructive and molecule specific technique. As discussed in section 2, Raman spectroscopy involves illuminating the sample with laser radiation with wavelengths either in the near-infrared (NIR), visible or ultraviolet (UV) regions, monitoring the light reflected from the sample and analyzing the intensities as a function of wavelength.

The focus of the chapter is to discuss the applicability of Raman spectroscopy as a non-destructive and molecule specific tool for monitoring food quality. This goal is achieved through a discussion of the basic properties of Raman scattering (RS) and experimental aspects, followed by a discussion of three case studies: 1. Revelation of a pork content in minced lamb products, 2. Detection and classification of nearly identical anti-oxidants and 3. Detection of pesticides on fruits and vegetables using surface enhanced Raman scattering (SERS).

In general the requirements to any experimental method suitable for an on-site evaluation of food quality are:

1. robust and easy to use instrumentation
2. portable instrument
3. non-destructive measurements
4. no or a minimum of sample preparation
5. a fast acquisition time
6. a qualitative and quantitative determination of chemical constituents
7. a high molecular specificity
8. a measurement of low concentrations of unwanted contents

Several of these requirements can be met by optical techniques based on some kind of reflection measurement. The importance of the different requirements will depend on the specific application, e.g. are the development and implementation of the technique highly dependant on, whether the commodity should be controlled for an unwanted or wanted content. Both types of content place requirements on the molecular specificity, however, in the case of detecting an unwanted content, the concentration is often very low as well.

Figure 1 illustrates the contents of information obtained from three different kinds of reflection measurements performed on the same green leaf.

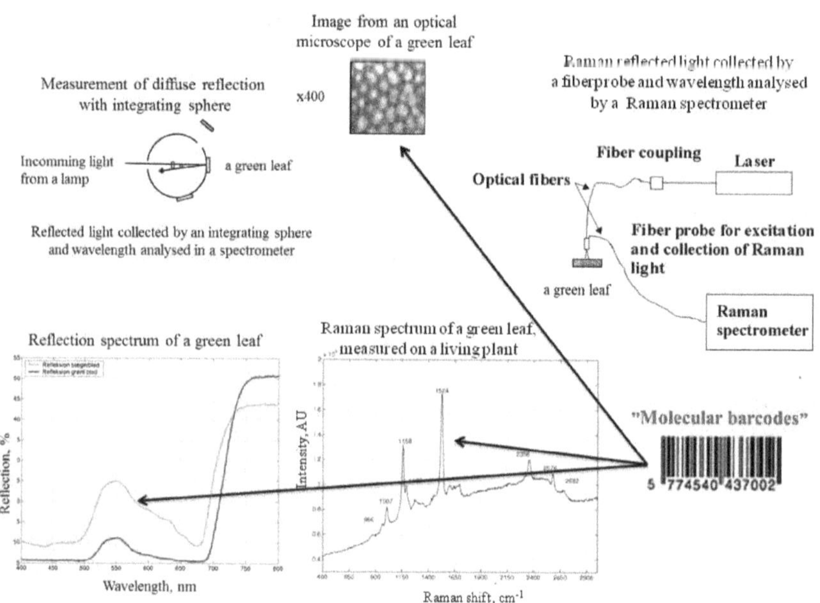

Fig. 1. Optical reflections with different information content from a green leaf. Left: Diffuse reflection (multiple scattering), middle: Diffuse reflection and imaging and right: Molecular reflection (Raman scattering).

The middle part of the figure shows an image obtained with an optical microscope, magnified 400 times, where the cells containing chlorophyll are resolved. This experimental method is based on diffuse reflection of white light and imaging and no quantification of the spectral information is made, except the information visible to the human eye. To the left is shown a diffuse reflection spectrum of the same leaf (red curve) obtained with a Perkin Elmer, λ900 spectrophotometer equipped with an integrating sphere. The integrating sphere collects all the light reflected from the leaf enabling an absolute measurement of the reflectance coefficient. Reflection measurements performed on different kind of green plants and on different plant parts are similar and contain almost identical spectral information. The example in figure 1 shows reflection spectra from two types of leaves of a Hibiscus, rosa sinensis, namely a foliage leaf and a sepal Jernshøj & Hassing (2009). The spectra shows differences in the absolute reflection values, but are very similar with respect to spectral information. The similarity is partly due to a blurring of the molecular signal caused by the molecular interaction with the surroundings and partly, since the molecular signal in itself is a composite signal reflecting both the motions of the electrons and nuclei in the molecule. A closer study shows a higher concentration of secondary pigments, e.g. carotenes, in the sepals than in the green leaves. A quantitative analysis of chlorophyll and carotenoid from these spectra is possible, when applying empirical models, this reflects the complex scattering and interaction processes taking place in the leaf Jernshøj & Hassing (2009); Kortüm (1969). However, due to the poorly resolved spectra, it may be impossible to discriminate between the presence of closely related molecular species, such as the antioxidants α- and β-carotene.

The right part of figure 1 shows a Raman spectrum obtained from the same foliage leaf. As opposed to the diffuse reflection measurements, the application of a laser results in the generation of a molecular reflection signal with measurable intensity, namely the Raman scattered light. As clearly seen from the figure, the spectral information is increased dramatically in the Raman spectrum. As discussed in the next section, the spectral distribution observed in the spectrum primarily reflects the vibrational motion of the nuclei in the "naked" molecules.

Summarizing: The outcome of an optical reflection measurement may be compared to a bar code, which is a well known component in different industries, where different and often high information content is encoded into this code and placed on a commodity. The information is read out by measuring the reflected laser light from the bar code, an example being the price scanner used in supermarkets. The difference between this bar code and the molecular information, is that the "molecular bar codes" are native parts of the sample, which are basically determined by the molecular composition. The informational quality of the particular "molecular bar code" obtainable is defined by the type of interrogative process used, e.g. imaging, diffuse reflection or Raman scattering.

2. Raman spectroscopy

The section gives an introduction to Raman scattering and point to the potential inherently present in the Raman effect with respect to obtain detailed molecular information. The section focuses on the theoretical and experimental challenges that have to be overcome in order to make different kinds of Raman techniques valuable diagnostic tools in the analysis of food quality.

Raman spectroscopy involves illuminating the sample with laser radiation with wavelengths either in the NIR, visible or UV regions, which excites the constituent molecules within the sample to vibrate. A vibrational Raman spectrum of the molecules is obtained by collecting the in-elastically scattered light. Each molecule present in the sample has a characteristic set of nuclear vibrations and thus the sample as a whole has a unique vibrational signature, i.e. a "molecular bar code" with a high information content. Raman spectroscopy is a class of well-documented, non-destructive, optical techniques with a high spectral resolution all of which are based on the Raman effect discovered by C. V. Raman in 1928 Raman & Krishnan (1928). Today more than 25 different Raman spectroscopies are known Long (2002).

2.1 An experimental view on Raman spectroscopy

Raman spectra can be obtained as reflectance measurements, which means that samples can be investigated with no or very little sample preparation and as opposed to other widely applied optical techniques, such as NIR and Fourier Transformed Infrared (FT-IR), Raman measurements are not influenced by the presence of water and therefore biological samples can be measured in their natural environment. Besides, the samples are not influenced by the measurements and the same samples can be investigated over time, which is essential, when measuring on food samples. Since the spectral information contained in a non-resonance Raman spectrum (vide infra) is virtually independent of the laser wavelength and since a complete Raman spectrum, typically 0 - 3500 cm^{-1}, only covers a wavelength region of approximately 100 nm (the exact value depends on the specific laser wavelength), it follows that complete Raman spectra of food products can be measured without removing the protective film covering the products, just by choosing a laser wavelength that matches the optical window of the protective film.

Raman spectrometers may be divided into two classes: Dispersive instruments and FT-Raman instruments. Any dispersive Raman spectrometer consists essentially of four components, a filter to block the Rayleigh scattered light, an entrance slit (often defined by an optical fiber), a transmission or reflection grating, where in the latter case the focusing optics is built into the grating and a CCD detector, which is coupled to a computer. The image of the illuminated entrance slit or fibre core is formed on the CCD and the different wavelengths contained in the Raman signal is converted by the grating into different positions on the CCD. Because of the simplicity of the basic Raman spectrometer, it is possible to build different editions for different purposes.

Figure 2a shows a typical Raman spectrometer, suited for scientific purposes. The setup, which is developed at The Molecular Sensing Engineering group, Faculty of Engineering, Institute of Technology and Innovation, University of Southern Denmark, has been built having a high degree of flexibility in mind. This flexibility allows us to arrange and rearrange the setup according to the experimental conditions necessary to achieve the desired molecular information. One of the main research areas is molecular investigations on bio-molecules, such as porphyrin and haemoglobin doing resonance and non-resonance Raman spectroscopy, polarized and unpolarized experiments as well as Surface Enhanced Resonance Raman Scattering (SE(R)RS) Jernshøj & Hassing (2010). Especially research involving polarized Raman measurements, where the molecular information obtained using linearly or circularly polarized light, has been carried out on a number of different samples. Recently, the molecular information obtained from such polarized measurements on a highly symmetric gold nanostructure (SE(R)RS) has been investigated in details K. D. Jernshøj & Krohne-Nielsen (2011).

When combining a Raman spectrometer with an optical microscope, the information content may be further increased. The Raman setup in figure 2a consists of a modified Olympus BX60F5 microscope, a SpectraPro 2500i spectrograph from Acton (Gratings: 1200 and 600 lines/mm) and a cooled CCD detector from Princeton Instruments, model Acton PIXIS. The setup is equipped with 12 different laser excitation wavelength, provided by: a 532 nm diode laser (Ventus LP 532), a Spectra Physics 632.8 nm HeNe laser, a Spectra Physics Ar^{+} laser: visible region and a tunable Ti3+:sapphire (titanium sapphire) laser: visible and NIR region. The setup has adjustable spectral, 5cm^{-1} and spatial resolutions, 0.3μm. Since the Raman spectrometer is combined with an optical microscope, equipped with a motorized, translational XY-stage (Thorlabs Inc., CRM 1) on the microscope translational stage, it is possible to obtain complete Raman spectra from different points across the sample. Due to the high spatial resolution (0.3μm) it is possible to perform Raman Imaging with a subcellular

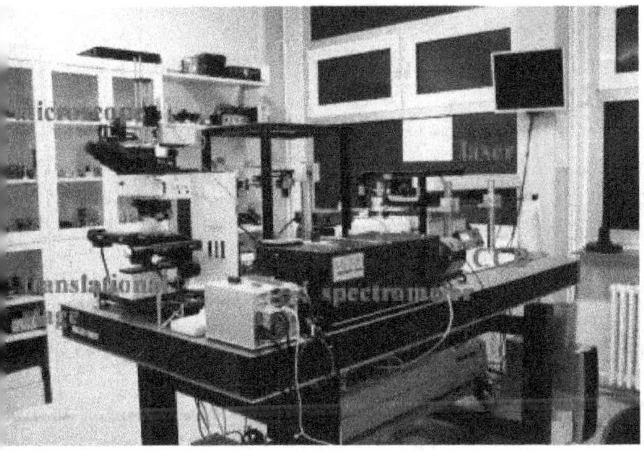

Fig. 2. The Raman equipment at The Molecular Sensing Engineering group at Faculty of Engineering, Institute of Technology and Innovation, University of Southern Denmark. (a.) The Raman setup, which is suited for scientific purposes. The Raman spectrometer is combined with an optical microscope, which is equipped with a motorized XY-stage on the translational stage. (b.) A portable commercial Raman setup: DeltaNu Inspector Raman, 785 nm excitation, unpolarized Raman, spectral resolution: 10, 12 or 15 cm^{-1}, predefined laser power 1.8, 3.4 and 6 mW, polystyrene standard for wavelength calibration, battery operated. Inspector Raman has been equipped with a manual translational XY-stage, order to be able to translate the laser across the sample.

spatial resolution, which has proven especially useful combined with multivariate analysis in the study related to early diagnosis of human breast cancer cells Martin Hedegaard & Popp (2010). As discussed in case study 3.1, Raman imaging combined with a machine vision system can be applied in an automatized determination of the composition of inhomogeneous food products, such as minced meat products.

For the purpose of an on-site, non-destructive investigation, the laboratory is also equipped with a commercial portable Raman spectrometer, which is shown in figure 2b. The parameters of this turn-key instrument are: 785 nm excitation, unpolarized Raman measurements, predefined spectral resolutions: 10, 12 or 15 cm^{-1}, predefined laser power 1.8, 3.4 and 6 mW, polystyrene standard for wavelength calibration and finally the possibility for battery operation. According to requirement 2 on page 2, dealing with portability, this kind of small Raman spectrometer is suited for field use, since it may be battery operated and is portable. Besides, the spectrometer has been equipped with a manual XY-stage in order to be able to scan the laser across the samples, when measurements are done on inhomogeneous samples.

As will be discussed in the next section, laser induced fluorescence, which is excited simultaneously with the Raman process, may often be a serious problem in practical applications of Raman spectroscopy. The fluorescence can be avoided in most cases by choosing the laser wavelength in the NIR region. When this choice is combined with an instrument that has a high sensitivity (high through-put), Raman spectra of high quality can be obtained. In FT-Raman spectrometers the grating is replaced by a scanning interferometer (e.g a Michelson interferometer) by which an interferogram, i.e. a time signal containing the spectral distribution of the Raman signal, is measured. The Raman signal is calculated by a computer by performing a fast Fourier transform of the interferogram data. A FT-Raman spectrometer consists of the following components: a NIR-laser with wavelength 1064 nm, filters to block the Rayleigh scattered light, a stable and efficient interferometer, a sensitive detector coupled to a computer, which includes software with the capability of performing fast Fourier transform. Typically, FT spectrometers are scientific instruments but it is possible to buy a FT-Raman spectrometer, e.g. RamanPro, which is designed to measure chemicals in production environments (http://www.rta.biz/).

Since different portable, commercial Raman instruments with different specifications are available, it will be possible to optimize the application with respect to the specific goal by applying e.g. either multivariate analysis, some form of enhancement of the Raman signal and specially developed additional software. Notice, that in the case of FT-Raman instruments the above mentioned flexibility with respect to choosing the laser wavelength is limited.

2.2 Fragments of Raman theory

As mentioned previously Raman spectroscopy involves illumination of the sample with laser radiation with wavelengths either in the NIR, visible or UV regions. According to quantum mechanics the intensity of a laser beam is proportional to the number of photons and the photon energy Long (1969). In molecular spectroscopy it is custom to measure the molecular and photon energy in the unit cm^{-1}, where x $cm^{-1} = 10^7/y$ nm. This means that the photon energy of a NIR laser with wavelength y = 785 nm is equal to x = 12739 cm^{-1}, while the photon energy for a visible laser with wavelength 532 nm is 18797 cm^{-1}. Thus, the photon energy is inversely proportional to the wavelength.

When the wavelength of the laser is chosen in the spectral region, where the molecules in the sample do not absorb any light, a fraction of the laser light is scattered. Most of the scattered light will deviate from the incident laser light only in the direction of propagation, but will have the same wavelength as the laser. This scattering process is known as Rayleigh scattering. A small fraction of the scattered light, namely the Raman scattered light, will in the scattering process also be shifted in wavelength. These measurable shifts are determined by the physical properties of the scattering molecule and they appear in the Raman spectrum as characteristic sharply defined peaks as seen in figure 1. Since each molecule gives rise to a characteristic and unique Raman spectrum, we are presented with a "molecular bar code" with a high information content. In vibrational Raman spectroscopy, the spectra reflect that each molecule has a characteristic set of nuclear vibrations. Besides, being shifted in wavelength the polarization of the Raman scattered light may also be different from the polarization of the laser light. The shift in polarization is determined by the nature of the various nuclear vibrational motions, which are also determined by the molecular properties. The quality of the information contained in the "molecular bar code" can therefore be increased by including the shift in polarization in the analysis.

In the following we focus on some of the basic principles describing the Raman process, which are necessary to illustrate the potential of and difficulties with Raman spectroscopy, when applied in food analysis. The fundamental theory, various aspects and applications of Raman scattering have been extensively discussed in the literature and we refer to these for further details eds. R. J. H. Clarke & Hester (n.d.); Long (2002); McCreery (2000); Mortensen & Hassing (1980); Plazek (1934); Smith & Dent (2005).

Figure 3 gives a schematic overview of the Raman process (left) and the fluorescence process (right) and the basic expression for the Raman intensity I_{Raman} is given in Eq.(1).

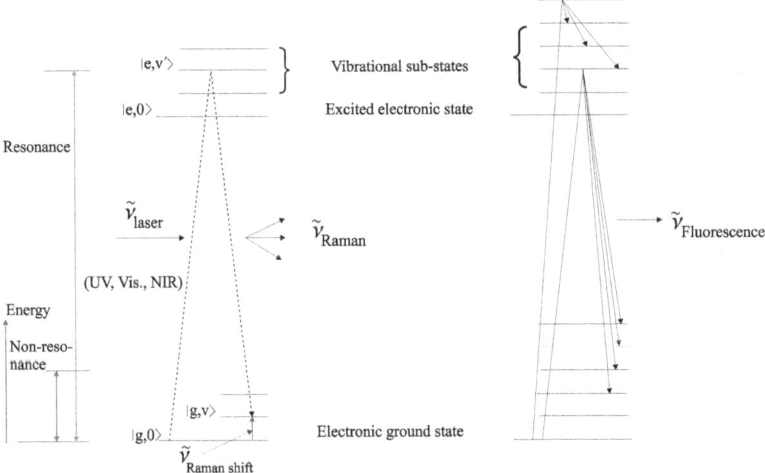

Fig. 3. Left: The Raman process is a *coherent* absorption-emission sequence. Right: Fluorescence is a real absorption followed by a spontaneous emission process, i.e. an *incoherent* absorption-emission sequence.

$$I_{Raman} \propto \tilde{\nu}^4_{Raman} \left| \sum_{e,v'} \frac{\langle gv|\rho|ev'\rangle\langle ev'|\sigma|g0\rangle}{\tilde{\nu}_{ev',g0} - \tilde{\nu}_{laser} - i\gamma_{e,v'}} + \frac{\langle gv|\sigma|ev'\rangle\langle ev'|\rho|g0\rangle}{\tilde{\nu}_{ev',gv} + \tilde{\nu}_{laser} + i\gamma_{e,v'}} \right|^2 \cdot I_{laser} \quad (1)$$

The horizontal bars represent the energy of the molecular states, 0 is the vibrational ground state (indicating that none of the molecular vibrations are excited) and v is the final vibrational state. $|g,0\rangle$ and $|g,v\rangle$ are the symbols for the initial and final molecular states participating in the Raman process, where g denotes the electronic ground state and 0 and v the vibrational substates of the initial and final states. $|e,v'\rangle$ denotes an electronic excited state in the molecule and its vibrational substate. $\rho,\sigma = x,y,z$, where $e \cdot \rho$ is the electric dipole moment (e is the charge of an electron). Since each molecule has a characteristic set of nuclear vibrations, v and v' are really sets of numbers, where the value of each number describes the degree of excitation of each kind of vibrational motion. Thus, $v = v_1, v_2, ...v_k, ...v_{3N-6}$, where N is the number of nuclei in the molecule. In the benzene molecule e.g., where $N = 12$, the set will consist of thirty numbers, corresponding to the thirty different kind of vibrations that may be excited.

The Raman process can be thought of as a *coherent* absorption - emission sequence, in which the absorption of the incoming laser light is followed by an immediate re-emission of the Raman scattered light. During the initial absorption the molecule shifts state from $|g,0\rangle$ to $|e,v'\rangle$, while during the re-emission the molecule shifts state from $|e,v'\rangle$ to $|g,v\rangle$. In non-resonance Raman Scattering, where the photon energy of the laser (illustrated by the red arrow in figure 3) is small compared to the energy of any electronically excited state, all molecular states, $|e,v'\rangle$, contribute to the scattering process and thereby to the intensity of a particular Raman line in the spectrum. This is reflected through the appearance of a summation over all molecular states of the molecule in the theoretical expression for the Raman scattered intensity given in Eq.(1).

For comparison, the fluorescence process is illustrated to the right in figure 3. This process is an *incoherent* absorption - emission sequence, which consists of a genuine absorption of incident light followed by a genuine emission of light. As illustrated in the figure, the initially excited molecule is allowed to shift state before it spontaneously emits light, which destroys the coherent nature of the sequence. When focussing on the emission spectrum (i.e. the fluorescence), the molecule may during the emission end up in different vibrational substates. Since the contributions from all the possible transitions have to be added together in the expression for the emission intensity and since the number of vibrational motions, $3N - 6$, may be very large for molecules typically appearing in food products, the different contributions to the intensity overlap with the result that the spectral distribution in the fluorescence spectra becomes broad and without much structure. This decreases the quality of the "molecular bar code" related to fluorescence. Depending on the experimental conditions, both the Raman and fluorescence processes may be initiated simultaneously. This is illustrated in the Raman spectrum of the green leaf in figure 1, where the Raman spectrum "rides" on top of a broad fluorescence background. The fluorescence may even be so pronounced that the Raman spectrum is partly or completely hidden. One way to avoid the excitation of fluorescence is to choose a laser with a photon energy smaller than any electronic excitation energy of the molecules.

It follows from the expression for the Raman intensity given in Eq.(1) that the intensity is proportional to the intensity of the laser and proportional to the fourth power of the photon energy of the Raman scattered light. The energy of the final molecular state termed $\tilde{\nu}_{Raman\ shift}$ is equal to $\tilde{\nu}_{Raman\ shift} = \tilde{\nu}_{laser} - \tilde{\nu}_{Raman}$. In Raman spectroscopy one measures I_{Raman} as a function of $\tilde{\nu}_{Raman\ shift}$, see figure 1. Since $\tilde{\nu}_{Raman\ shift}$ is equal to a vibrational energy of the molecule, the Raman spectrum depicts the characteristic vibrations of the molecule. Vibrational energies are typically much smaller than the excited electronic energies of a molecule, this means that the Raman intensity is approximately proportional to the fourth power of the photon energy of the laser. Experience shows that the fluorescence may in most cases be avoided by choosing a NIR laser with wavelength 1064 nm, which corresponds to the photon energy: 9399 cm^{-1}. Comparing the latter with the photon energy of the visible laser with wavelength 532 nm, the Raman intensity is reduced by a factor 16. The loss of Raman intensity is in the FT-Raman spectrometers mainly compensated for by the removal of fluorescence combined with the high sensitivity that can be obtained in these instruments Vidi (2003).

The central part of the expression is the absolute square of the molecular polarizability, $\alpha_{\rho\sigma}$, where $\alpha_{\rho\sigma}$ describes the change in the electron distribution of the molecule in response to the interaction with the incoming laser light. Each numerator in $\alpha_{\rho\sigma}$ contains a product of two terms, where each term is called the transition dipole moment, reflecting the transitions taking place in the molecule during the scattering process. The magnitude and sign of these in combination with the magnitude and sign of the denominator determine the contribution from a specific molecular state to the intensity of a particular Raman line. The appearance of the absolute square of the sum of these contributions reflects the coherent nature of the Raman process. In fluorescence both the absorption and the re-emission processes are independent and determined by the absolute square of each transition moment. *In fact, the higher information content observed in Raman spectroscopy has origin in the coherent nature of the Raman process.* The real part of the denominator in the first term, called the resonance term, contains the energy difference between the energy of the excited state $|e, v'\rangle$ and the photon energy of the laser, $\tilde{\nu}_{ev',g0} - \tilde{\nu}_{laser}$. The imaginary term reflects the energy broadening of the state $|e, v'\rangle$. If the photon energy of the laser is chosen equal to $\tilde{\nu}_{ev',g0}$, the contribution from this state to the Raman process becomes dominating causing the Raman signal to be enhanced. The enhancement depends on the value of the imaginary term but the enhancement may typically be of the order of magnitude 10^4 to 10^6. The phenomenon is termed Resonance Raman Scattering (RRS) and is illustrated with the green arrow in figure 3. In practice the molecular states close to the resonant state will give the largest contribution to Raman scattering. In RRS it is possible to get information about the participating excited states, whereas in non-resonance this information is smeared out.

A large bio-molecule, e.g. present in various food products, has a large number of characteristic vibrations. Although, not all of them are seen in the Raman spectra, the spectra may often be very complex, i.e. the "molecular bar code" contains a lot of information. Raman scattering can be performed in several ways with the result that specific parts of the obtainable molecular information may be reached, i.e. distinct parts of the "molecular bar code" is read out. Resonance enhancement of the Raman signal can be utilized to obtain information about specific parts of a large molecule, e.g. if the molecule contains a chromophore, it is possible to select the laser wavelength close to an electronic absorption of the chromophore with the result that only those vibrations involving the nuclei of the chromophore are enhanced and therefore seen in the Raman spectrum. A chromophore is the part of a molecule responsible for its color, where the color arises, when the molecule absorbs certain wavelengths of visible light and transmits or reflects others.

A special kind of resonance Raman spectroscopy, termed Raman Dispersion Spectroscopy (RADIS) Mortensen (1981), involves a quantitative comparison of Raman spectra measured with a few different excitation wavelengths close to the electronic resonance of the chromophore. The 3D graph in figure 4 shows the possibilities with RADIS. For each excitation wavelength the corresponding Raman intensity can be followed as a function of the Raman shift, while when choosing a specific Raman shift, the intensity of this can be followed as a function of the excitation wavelength. Due to the narrow line width of the Raman spectra and the coherent nature of the Raman process each Raman band has its own and distinct excitation spectrum (termed an excitation profile).

As discussed in subsection 3.2 this enables discrimination between molecularly almost identical constituents, such as α- and β-carotene even when the amount of one of the constituent is very small compared to the other.

2.3 Improving Raman sensitivity by nanotechnology

Raman spectroscopy has two serious limitations: First, Raman scattering is inherently a weak effect (typically 10^8 incoming laser photons only generate 1 Raman photon) and secondly, fluorescence is often emitted concurrently with the Raman scattering. Since the fluorescence signal is typically 4 to 8 orders magnitude larger than the Raman signal, this will be hidden in the fluorescence background. The fluorescence may stem from the molecules under investigation or from other molecules in the sample. The latter situation may often arise when measuring on food samples, where the sample preparation is absent or kept at a minimum. The weakness of the Raman signal may be improved in different ways. (1): by increasing the photon energy of the laser (i.e. by choosing a laser with shorter wavelength) so that it corresponds to a resonance region of the molecule, (2): by improving the signal to noise ratio

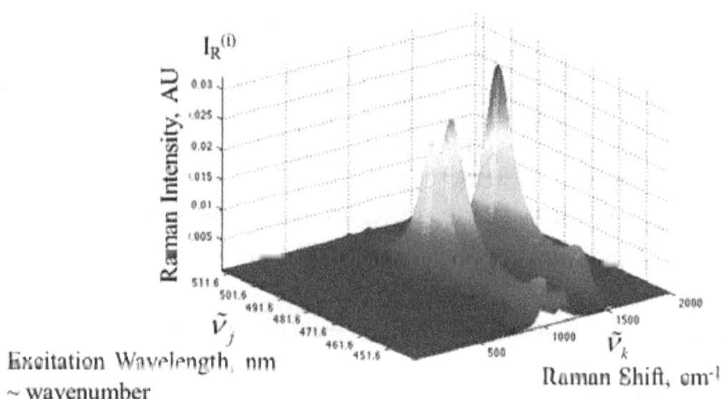

Fig. 4. RADIS, Raman spectra measured with a few different excitation wavelengths close to the electronic resonance of the chromophore.

through the application of a sensitive and cooled CCD detector and (3): by the application of advanced signal processing and multivariate methods. Although, a resonance enhancement of the Raman signal can be achieved by increasing the photon energy of the laser, the amount of absorption is also increased, which may lead to photoinduced degradation of the sample. Since in general the amount of fluorescence increases with shorter wavelength, the resonance enhanced Raman signal may still be partly hidden in the fluorescence background.

On the other hand, when the sample is exposed to a laser with longer wavelength the fluorescence decreases, but unfortunately, this also leads to a lower Raman intensity through the $\tilde{\nu}^4_{Raman}$ dependence shown in Eq.(1). Even though the choice of a longer laser wavelength may result in an acceptable Raman signal, it would be advantageous to enhance the Raman signal and in particular relative to the fluorescence.

This may be achieved in Surface Enhanced Raman Spectroscopy (SERS). The enhancement of the Raman signal may occur, when the scattering molecules are either physisorbed or chemisorbed to a nanostructured metallic surface often made of gold (Au) or silver (Ag). Although a large enhancement can be achieved (up to 10^{14} has been reported), the introduction of a nanostructured surface will in general influence the Raman signal from the native molecules and make the scattering process more complex, so that the interpretation and implementation of SERS require a detailed analysis of the system under investigation.

When a bare nanostructured metal surface is illuminated by the laser, the laser photons interact with the electrons in the surface layer. When the metal and the adsorbed molecules are exposed to laser radiation the incoming laser photons interact with the combined system with the overall result that the intensity of the Raman scattered light is enhanced relative to the intensity of the Raman signal of the free molecules and relative to the fluorescence. The interaction between the incoming light, the metal and the adsorbed molecules depends in a complicated way on the surface morphology, the kind of metal and on the molecule in question. A detailed discussion of the theory and the various implications of SERS can be found in the literature Jernshøj & Hassing (2010); K. Kneipp & (Eds.); Ru & Etchegoin (2009); Willets & Duyne (n.d.). Below, only a brief discussion of SERS is given. Figure 5 illustrates Raman scattering of molecules adsorbed to a nano-structured metal surface.

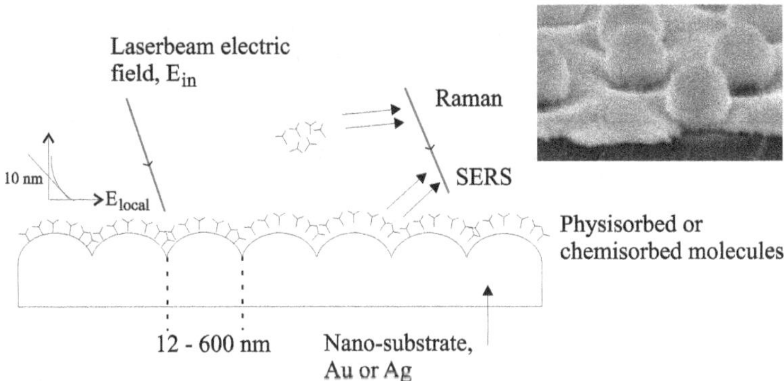

Fig. 5. A simplified schematic representation of SERS. The insert is obtained from R. L. Eriksen and O. Albrektsen (Faculty of Engineering, Institute of Technology and Innovation, University of Southern Denmark). Similar structures can be found in reference R. L. Eriksen & Albrektsen (2010).

A focused laser beam illuminates the molecules adsorbed to the Au or Ag surface. The diameter of the laser beam is typically 0.6 μm and the size of the nano-structure is 12 - 600 nm. The insert shows a SEM-micrograph of a real nano-substrate based on a Si-substrate with 100 nm spheres coated with a 40 nm thick layer of Au. Due to the interaction between the laser light and the electrons in the metal surface a local electric field E_{local} is created outside the metal and very close to the surface. Depending on the specific surface structure E_{local} may be much larger than the incoming electric field, E_{in}, associated with the photons in the laser beam and E_{local} may be rather different in different points on the surface. Surface points, where E_{local} is very high, are called hot-spots. As indicated E_{local} will decrease exponentially with the distance from the surface and only those molecules that are within approximately 10 nm will be influenced significantly by this field. According to electromagnetic theory, the intensity of a light wave is proportional to the absolute square of the electric field associated with the wave. As indicated in the figure, the incoming laser light gives rise to ordinary Raman scattering by the molecules, which are not close to the surface. The Raman intensity of these molecules is proportional to $| E_{in} |^2$. The molecules adsorbed at the surface interact with the local field and with the electrons in the metal. The result is that both the absorption and the emission parts of the Raman process become proportional to $| E_{local} |^2$, so that the SERS-signal becomes proportional to $| E_{local} |^4$. The enhancement of the SERS signal is therefore given by the factor $| \frac{E_{local}}{E_{in}} |^4$.

Thus in cases, where the local field is just $10E_{in}$, the enhancement of the Raman signal becomes equal to 10000. For comparison, the fluorescence signal is only enhanced with the factor $| E_{local}/E_{in} |^2$. The difference between the enhancement of the Raman signal and the fluorescence may be attributed to the difference in the coherence properties of the two processes.

SERS can be performed by using nano-substrates with an ordered structure of the SERS-active sites. These can be designed and fabricated with specific applications in mind, including functionalization of the surface with a layer of molecules, which bind reversibly to the specific molecules to be investigated. There are a large variety of commercial nano-substrates available on the market. Another possibility is to form metal colloids and mix the samples with the colloid solution or by coating e.g. a SiO_2 surface with Au aggregates. Colloidal Au has e.g. been used to chemically identify important components in plant material such as green tea leaves, shredded carrots or shredded red cabbage Zeiri (2007). SERS active substrates made of aggregated Au nanoparticles on a SiO_2 substrates have been applied to detect single cells of different bacteria, which are very important in relation to food products, i.e. Escherichia coli and Salmonella typhimurium W. R. Premasiri & Ziegler (2005). Although Au and Ag based substrates or colloids are the most commonly used, other materials have been applied. Thus, the SERS signal from single cells of Escherichia Coli bacteria has been obtained by mixing ZnO nanoparticles with the bacteria cells R. K. Dutta & Pandey (2009).

2.4 Extracting information from "molecular bar codes" by multivariate analysis

Multivariate analysis is in general applied in order to analyze experimental data by the use of mathematical and statistical methods. The application of multivariate analysis to spectroscopic data has indeed become very important for describing small differences between chemical constituents in samples containing bio-molecules. Spectroscopic data are often analyzed by using Principal Component Analysis (PCA), to which a brief introduction will be given in the following, a thorough explanation of this and other multivariate methods can be found in reference A. K. Smilde & Geladi (2004).

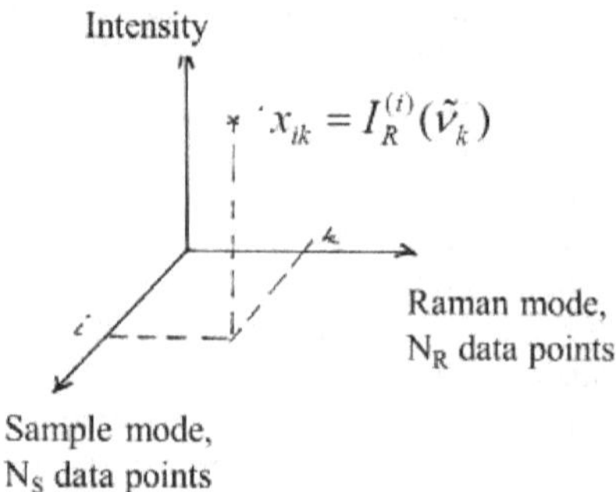

Fig. 6. Raman data matrix: A graphical representation of the data set obtained from N_S Raman experiments, where i is the sample number, $I^{(i)}_{Raman}(\tilde{\nu}_k)$ is the Raman intensity at the k'th energy position in the Raman spectrum.

Assuming that the Raman data from an experiment involving several samples are collected in a matrix denoted $\mathbf{X}$, where the matrix element x_{ik} represents the measured Raman intensity for the i'th sample at the energy $\tilde{\nu}_k$ in the Raman spectrum. The number of samples are termed N_S, where $i = 1, 2, ..., N_S$ and each Raman spectrum consists of N_R data points, where $k = 1, 2, ..., N_R$. The data set, containing the Raman intensities for all samples, will then represent a 2-way multivariate data set, which has the dimensions $N_S \times N_R$. A graphical representation of the above is shown in figure 6.

Another way to represent these data is described as follows. The N_R data points define a N_R dimensional coordinate system, where the axes are defined by the energies $\tilde{\nu}_k$. The Raman spectrum for the i'th sample is then represented by a single point, where the position of the point is determined by the Raman intensities at the different energies of the Raman spectrum. The distribution of the N_S sample points reflects the systematic differences between the Raman spectra of the individual samples. This means that sample points, which represent samples with similar Raman spectra, will lie close together. In general the dimension N_R of this coordinate system is large and some of these dimensions account for similarities in the Raman spectra, i.e. no Raman signal (noise) or identical features in the spectra.

The overall principle behind PCA is to make a favorable coordinate transformation, which represents the significant intensity variations in the spectra. The transformation is in principle carried out by defining a new rotated coordinate system of dimension N_C, where N_C is the number of principal components and typically it is found that $N_C \ll N_R$. For that, a step-by-step procedure is applied. If we assume that the same spectroscopic structure (not necessarily with the same absolute intensity) is present in the Raman spectra in the majority of the samples, then the distribution of the sample points define an average direction in the N_R coordinate system. The unit vector defining this direction is termed the loading vector $\vec{p}_1$, which is calculated from the data by least squares minimization. The score, t_{i1}, for the i'th sample point, is defined as the coordinate of this point in the direction of the loading vector, $\vec{p}_1$. The first principal component is calculated as the outer product (denoted $\otimes$) between the score and the loading vector, and hence is a matrix. This process is carried out again defining next a second loading and second score vector, the process is repeated until the desired accuracy is obtained. In Eq.(2) is given the mathematical expression for the data matrix Hedegaard & Hassing (2008),

$$x_{ik} = \sum_{c=1}^{N_c} t_{ic} p_{kc} + e_{ik}^{(N_C)} \tag{2}$$

where c is the principal component index and $e_{ik}^{(N_C)}$ are elements in a matrix, termed the residual matrix, $\boldsymbol{E}^{(N_C)}$, which in the ideal case contains only noise.

The information obtained from the PCA may be visualized in different ways. One may plot the loadings $\vec{p}_{kc}$ as a function of the energy index, k, for $c = 1,2,3,..., N_C$. The plot of the c'th loading is proportional to the average Raman spectrum related to the c'th principal component. If the Raman spectra for the N_S samples are almost identical, the plot of the first loadings will show an average of the common features of the spectra, whereas the manifestation of the differences happens in loadings of higher order. Another possibility is to plot the scores, t_{ic}, as a function of the sample index, i for $c = 1,2,3,..., N_C$, which illustrates the contribution to each loading from each sample. A third option is the score plot, where two of the N_c scores are plotted against each other. The score plot will show a grouping of the sample points according to how significant each of the two loadings contribute to the multivariate model. Since the goal of chemical classification problems is to find the eventual minor differences between the Raman spectra of the different samples, the two latter type of plots clearly reveal, which of the samples are classified correctly as well as the uncertainty of the classification.

The coherent properties of the Raman process can be utilized to create a 3-way multivariate data set. For each sample, the data matrix is constructed by measuring the RADIS data as depicted in figure 6. The elements of the RADIS data matrix x_{ijk}, where the additional index j defines the photon energy of the laser, are $x_{ijk} = I_R^{(i)}(\tilde{\nu}_j \tilde{\nu}_k)$. The 2-way PCA model, which was discussed above, may be generalized into a 3-way model, called Tucker 3 A. K. Smilde & Geladi (2004). A discussion of the mathematical details of applying Tucker 3 on various kind of spectroscopic data can be found in A. K. Smilde & Geladi (2004). Tucker 3 has proven most useful, when compared to other 3-way models, for treating Raman based data A. K. Smilde

& Geladi (2004); Hedegaard & Hassing (2008), which is caused by the high quality of the "molecular bar codes" produced in both dimensions of the RADIS data matrix. In the Tucker 3 model, the elements of the data matrix can be expressed as:

$$x_{ijk} = \sum_{a=1}^{N_A} \sum_{b=1}^{N_B} \sum_{c=1}^{N_c} a_{ia} b_{jb} c_{kc} g_{abc} + e_{ijk}^{(N_A N_B N_C)} \tag{3}$$

where $e_{ijk}^{(N_A N_B N_C)}$ are elements in the residual matrix, a_{ia} are elements in the score vector, b_{jb}, c_{kc} are elements in the loading vectors corresponding to the two dimensions defined by the excitation energy and the Raman shift, and g_{abc} are elements in a matrix reflecting the interactions between the various scores and loadings. One "Tucker 3 component", $\vec{a}_a \otimes \vec{b}_b \otimes \vec{c}_c \cdot g_{abc}$, is analogue to one principal component in PCA, $\vec{t}_c \otimes \vec{p}_c$.

In the 2-way PCA model, described by equation 2, the scores and loadings are determined in the 2 dimensions so that the data is represented by fewest possible parameters. In the 3-way case, the goal is basically the same, but now it is necessary to decompose the three dimensions simultaneously. It is therefore necessary to calculate two loading vectors describing both the Raman and the RADIS spectra for each score vector along the sample dimension of the data. A Tucker 3 model will be applied in subsection 3.2 in the discrimination between molecularly almost identical constituents, α- and β-carotene, when the amount of one of the constituent is very small compared to the other.

3. Three real life applications of Raman spectroscopy and multivariate analysis

The applicability of Raman spectroscopy in food analysis is demonstrated through the discussion of three real life applications, namely 1.) Revelation of a content of minced pork in minced lamb by applying Raman imaging and multivariate analysis, 2.) Discrimination between two anti-oxidants in a mixture by applying RADIS and three-way multivariate analysis and 3.) Non-destructive detection of chlorinated pesticide residues on fruits and vegetables applying a portable Raman spectrometer and SERS.

3.1 Raman imaging and multivariate analysis in the revelation of a content of pork in minced lamb

In the first application, it is demonstrated that Raman spectroscopy can be applied to identify a content of pork in minced lamb products available in e.g. supermarkets. The vision is to develop a scanning system operated by the consumer, which automatically, fast and non-destructively controls for a specific content, e.g. pork and reports the outcome to the consumer. A minced lamb product constitutes an inhomogeneous sample, which consists of areas of meat and fat. The scanning system should be based on a combination of Raman measurements and a multivariate analysis, where the laser excitation wavelength has been chosen in accordance with the transmission spectrum of the film covering the minced meat tub. The discussion of the application is divided into two parts, part 1. Methodology and part 2. Implementation. In part 1, the molecular marker(s) in the Raman spectra must be identified, which enables a distinction between meat and fat. The first part is initiated by using the Raman imaging system (532 and 632,8 nm excitation) shown in figure 2a to create a reference Raman data set of isolated fat and meat from pork and lamb. The XY scanning facility of the Raman imaging system is used in order to automatically create a statistically large reference data set. Measurements have been carried out changing the wavelength and varying the laser power in order to judge, which parameters gave the measurements best suited for identifying the markers and the multivariate analysis. Since the focal distance is changed during the measurements a software has been developed in order to maximize the Raman data by selecting the acceptable Raman spectra from the raw data. The change of focal distance has to be considered in a practical implementation of the method. The different measurements have besides been optimized with respect to a correction for the background signal and normalization. It is found that the spectral region, containing the most information useable in the identification of molecular markers, is from 500 to 1700 cm^{-1}. One conclusion to be drawn from the above measurements is that the wavelength 632,8 nm is more suitable for further measurements, since the laser wavelength 532 nm is placed in a region of larger absorption and hence may lead to sample degradation.

Figure 7 shows averaged reference Raman spectra of freshly slaughtered pork and lamb (meat and fat) obtained with the Raman imaging system, 632,8 nm excitation and the laser power 11 mW on the sample. The numbers given in the brackets are the total numbers of spectra, which are acquired on different places of a given sample.

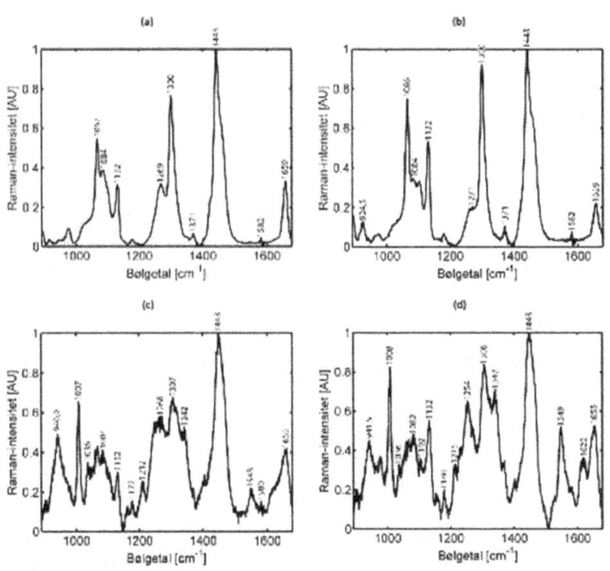

Average spectra : (a) pork-fat, (b) lamb-fat, (c) pork-meat , (d) lamb-meat

Fig. 7. Averaged reference Raman spectra (632,8 nm excitation) from (a.) pork fat (283 spectra), (b.) lamb fat (1493 spectra), (c.) pork meat (48 spectra) and (d.) lamb meat (239 spectra).

The method has been validated by measuring the Raman spectra from samples containing 100% minced pork (meat and fat) and 100% minced lamb (meat and fat). The relative amount of fat in these samples was not specified, but it was less than 20 percent. The number of spectra acquired for minced pork is 366 and for minced lamb 1320. The multivariate method is based on a Partial Least Squares - Discriminant Analysis (PLS-DA) A. K. Smilde & Geladi (2004), which, contrary to PCA, is a supervised method, where the data is classified according to predefined classes, in the present case the four classes: 100% meat and fat from pork and 100% meat and fat from lamb. The application of the PLS-DA is combined with an algorithm, which is illustrated by the flowchart in figure 8. The results of the validation are shown in table 1.

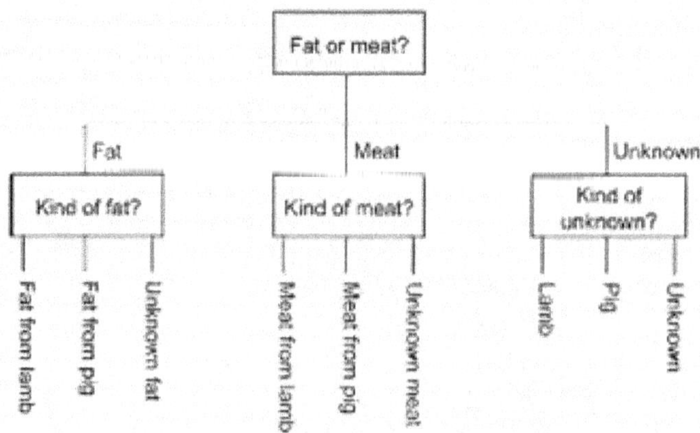

Fig. 8. A flowchart illustrating the different steps of the PLS-DA analysis.

Table 1 should be read as follows: The first row in the table corresponds to the first question in the flowchart and the columns correspond to the questions posed secondly in the flowchart. It follows from the table that the recognition ratio for 1. fat from lamb is 98.0 %, 2. meat from lamb is 20 %, 3. fat from pork is 89.0 % and 4. meat from pork 0 %. Notice that no sample from either lamb or pork has been classified wrongly, i.e. interchanged. The following percentages of the samples have been classified as unknown: fat from lamb fat 2.0 %, meat from lamb 80 %, fat from pork 11.0 % and meat from pork 100 %. It follows from the results that the discrimination between minced pork and lamb can be based on the Raman spectra of fat alone.

		Predicted Content				
Sample	*Content*	*Type*	*Quantity*	*Fat*	*Meat*	*Unknown*
1		Lamb	1257	724	10	523
	100% Minced Lamb	Pig	0	0	0	0
		Unknown	63	15	40	0
2		Lamb	0	0	0	0
	100% Minced Pig	Pig	313	299	0	14
		Unknown	53	37	13	3

Table 1. In the table are listed the results from a programmed algorithm for discrimination between minced lamb (1320 spectra) and pork (366 spectra). The Raman spectra from different points on the two kind of samples are obtained by an automatized XY scanning and 632.8 nm excitation (11 mW) using the Raman imaging system shown in figure 2a.

Further information about the experiments and multivariate analysis can be obtained from the authors (sh@iti.sdu.dk).

2. Implementation. A practical system based on a commercial, portable Raman instrument using 785 nm is under development. The implementation requires that one must consider the following: a.) localization of the areas of fat by using a machine vision system, b.) automatized scanning in order to obtain Raman spectra from fat, c.) automatic adjustment of focus d.) optimization of the excitation volume and laser power, e.) optimization of the statistics (the scanning area, the relative fat content in minced lamb and pork, respectively and the determination of the lower limit with respect to revealing a content of pork in a minced lamb product).

A further development will include the examination of other types of meat than pork and lamb, investigation of samples subjected to different pre-treatments such as freezing or storing inside or outside of a refrigerator. Research on these matters are in progress by the authors of the present paper and in reference Herrero (n.d.). The monitoring of meat quality with respect to molecular composition as a result of decomposition over time (ageing) applying a non-invasive and mobile system based on Raman spectroscopy and fluorescence has been performed in reference G. Jordan (n.d.). Besides, a micro-system including micro-optics and a compact external laser diode cavity with emission wavelength 671 nm suitable for Raman spectroscopy has been developed. The overall dimensions of the micro-system light source is $13\times4\times1$ mm^3.

3.2 Discrimination between two nearly identical anti-oxidants by applying RADIS and a Tucker 3 multivariate model

Carotenoids are a large family of pigments divided into two main groups carotene and xanthophyll, over 600 different pigments exist. Many of the carotenoid pigments are ubiquitous in nature and has attracted great interest in health and food science due to their nutritional importance Coultate (2002); H. D. Belitz & Schieberle (2004); K. Davies (2004). This importance also most importantly covers the composition of carotenoids in food products rather than specific individual pigments, this topic is subject to an ongoing discussion. This highlights the importance of being able to distinguish between very closely related molecular species, such as lutein, α- and β-carotene.

The second application is a chemical classification problem, in which one must discriminate between pure β-carotene and a mixture of α- and β-carotene. The challenge is that α- and β-carotene have nearly identical Raman spectra.

In reference H. Schulz & Baranski (2005), NIR-FT Raman spectroscopy has been applied in-situ in the analysis of intact plant material, in which the carotenoids are present in their natural concentrations. The carotenoids are present in a large variety of vegetables and fruits: orange, carrot roots, red tomato fruits, green French bean pods, broccoli inflorescence, orange pumpkin, corn and red pepper as well as nectarine, apricot, and watermelon. The spatial distribution of some carotenoids has been obtained by 2D Raman imaging. Although, the Raman spectra and images were recorded with a research, laboratory instrument (with 1064 nm excitation and spectral resolution 4 cm^{-1}), the same measurements could probably also be obtained with one of the recently commercially available portable FT-IR instruments combined with XY translational stage. The analysis performed in reference H. Schulz & Baranski (2005) is based on the three most intense Raman bands of the carotenoids around

1500 (ν_1 band), 1150 (ν_2 band) and 1000 (ν_3 band) cm^{-1}, also shown in figure 1. These bands are characteristic for all carotenoids, but depending on the specific carotenoid molecule, small vibrational shifts of the mentioned bands will be observed. The Raman spectra of the pure carotenoid standards, β-carotene, α-carotene and lutein can be found in figure 3 in reference H. Schulz & Baranski (2005).

In the analysis performed in H. Schulz & Baranski (2005) regarding anti-oxidants in tomatoes, the presence of the 1510 cm^{-1} Raman band has been interpreted as lycopen. A closer examination of the 1510 cm^{-1} band shows a small a-symmetry, which may indicate that β-carotene, where ν_1 = 1515 cm^{-1}, is also present but in a smaller concentration. Additional studies verify this interpretation. If the molecular species that contribute to the spectra are more similar than β-carotene and lycopen, the shifts of the corresponding Raman bands would be smaller and comparable to or less than the resolution of a portable spectrometer. Furthermore, if the difference in carotenoid concentrations is larger, it may be difficult or even impossible to solve a classification problem based exclusively on the vibrational shifts. The spectral distribution in the visible absorption spectra of most carotenoids are rather similar, however small variations in their color are observed. The change in color reflects small differences of the energy of the excited electronic states.

In the following we demonstrate how a small shift in electronic absorption energy can be utilized in a multivariate analysis of the vibrational Raman data. We consider a classification problem, in which the goal is to discriminate between samples with pure β-carotene and samples with a mixture of α- and β-carotene, where the concentration of α-carotene is very small compared to the concentration of β-carotene. From the Raman spectra of the carotenoid standards, it follows that α- and β-carotene have nearly identical Raman spectra: 1515 cm^{-1} (1521), 1156 cm^{-1} (1157) and 1007 cm^{-1} (1006), where the numbers in the brackets correspond to α-carotene.

The classification is performed by applying a Tucker 3 multivariate analysis to the RADIS data matrix obtained with the laser wavelengths 476.5, 488 and 497 nm. When comparing the laser wavelengths with the maxima of the absorption spectra of α- and β-carotene in references (n.d.a); H. D. Belitz & Schieberle (2004); Miller (1934), it is seen that we excite the carotene molecules close to resonance with the lowest electronic transition. The shift in electronic energy can be estimated from the absorption spectra to 268 cm^{-1}. *In the RADIS spectra, this shift will manifest itself as a different resonance enhancement of the Raman signal of the two pigments for each laser wavelength.* The RADIS data matrix will therefore from sample to sample vary along the two dimensions defined by the excitation energy and the Raman shift.

The experimental Raman and RADIS data are obtained from 10 samples, where 5 samples (1 - 5) contain only β-carotene and 5 samples (6 - 10) contain a mixture of 10% α-carotene and 90% β-carotene. The Raman spectra from one sample (476.5 nm excitation) of β-carotene (dashed line) and the mixture of α-carotene and β-carotene (solid line) in solution are shown in figure 9.

The results of a PCA and Tucker 3 analysis of the data are shown in figure 10.

Details about performing the Tucker analysis on similar RADIS data are given in Hedegaard & Hassing (2008). In both the PCA and in the Tucker 3 analysis the classification succeeds and the results are comparable, since the ratios between the average distance between the two classes and the variation within each class are similar. However, further studies involving

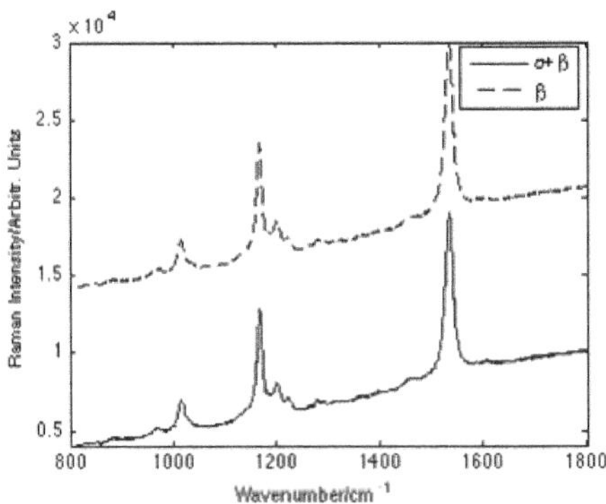

Fig. 9. The Raman spectra from one sample (476.5 nm excitation) of β-carotene (dashed line) and the mixture of α-carotene and β-carotene (solid line) in solution.

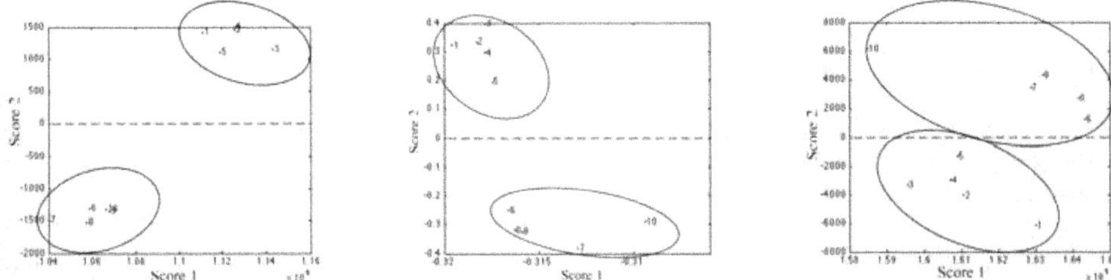

Fig. 10. (a.) Score plot of a 2 component PCA model, excitation wavelength 476 nm (score 2 (t_{i2}) is plotted against score 1 (t_{i1}) for $i = 1, 2....10$), (b.) Score plot of a 4 component Tucker 3 model, 476.5 and 496.5 nm (score 2 (a_{i2}) is plotted against score 1 (a_{i1})) and (c.) Score plot of a 2 component PCA model, 496.5 nm.

10 samples (1 - 10) with only β-carotene and 10 samples (11 - 20) with a mixture of only 0.5% α-carotene and 99.5% β-carotene show that the PCA in most cases leads to a wrong classification, whereas the Tucker 3 model in most cases still leads to a correct classification.

The robustness of the PCA and Tucker 3 models with respect to experimental uncertainties, such as fluctuations in the laser intensity, is different. This is demonstrated in the results shown in figure 11.

The number of samples and the concentration ratio are the same as in figure 10. In general one would expect the classification based on a Tucker 3 analysis to improve by incorporating an extra wavelength. However, by comparing figure 11b with 10b, it is seen that the classification succeeds, but the spread within the classes has increased. The result of the PCA analysis in 11a shows that the PCA classification fails, since it leads to a false grouping of the samples. A closer study of the experimental conditions reveals that the laser intensity at 488 nm fluctuates, which is the explanation for the above results.

In many cases the PCA or PLS-DA analysis of Raman data works fine due to the well resolved spectra. However, it is, as demonstrated above, possible to improve the analysis by including a shift in electronic absorption energy. Another possibility is to include the polarization properties of the Raman spectra in the analysis S. Hassing (2011).

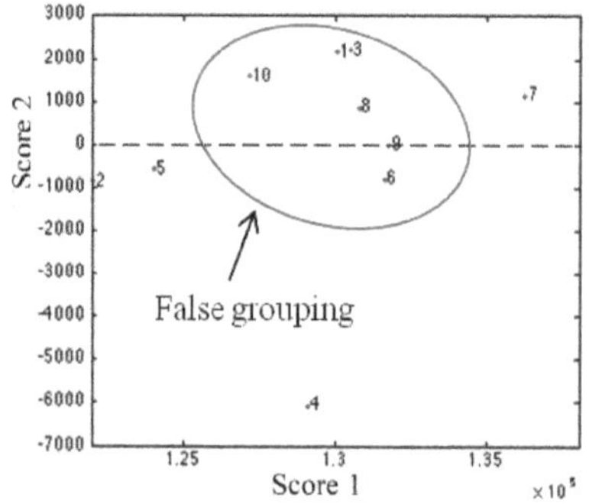

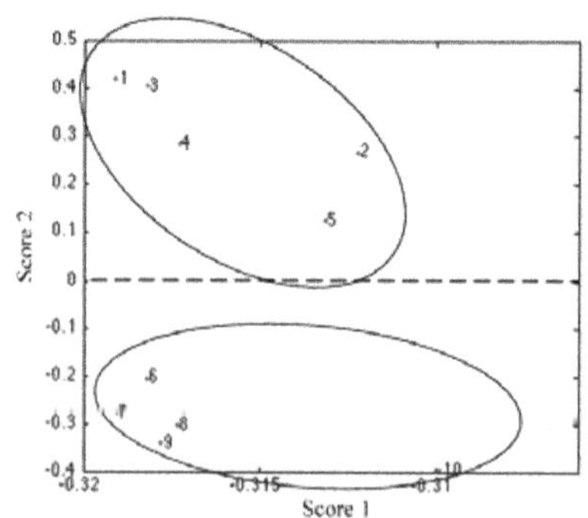

Fig. 11. (a.) Score plot of a 2 component PCA model, excitation wavelength 400 nm (score 2 (t_{i2}) is plotted against score 1 (t_{i1}) for $i = 1, 2....10$), (b.) Score plot of a 4 component Tucker 3 model, 476.5, 488 and 496.5 nm (score 2 (a_{i2}) is plotted against score 1 (a_{i1}) for $i = 1, 2....10$).

3.3 Non-destructive detection of chlorinated pesticide residues on fruits and vegetables applying a portable Raman spectrometer and SERS

We are today facing an increasing exposure to a cocktail of pesticides and other chemicals. The daily exposure to these chemicals, often used in horticulture, agriculture etc., is suspected to cause issues in human health, these include several diseases, reduced fertility and birth defects. One contribution to this exposure is the daily intake of pesticides through food consumption. Besides, an increasing concern with respect to the "cocktail effect", arising when digesting food products containing several pesticides, is seen, since the joint action is not fully understood Danish Ministry of the Environment (2005). Today, more than 800 different pesticides are available and a large number of those are applied to food products Danish Environmental Protection Agency (2005); Veterinary et al. (2003; 2005). Gas Chromatography (GC-multi method, FP017) has been the method of choice for such determination, this procedure, however, is both labor intensive, destructive and non-portable Veterinary & Administration (2003). A calculated intake of pesticides is found by multiplying the average consumption of a commodity with the content of the pesticide and a Hazardous Quotient (HQ) is found by dividing the above mentioned intake with the Acceptable Daily Intake (ADI). The most frequently found pesticides were not including those contributing the most to the hazard quotient, but e.g. dicofol.

It should be noticed that although dicofol is toxic, building up sediments in plants and animals and therefore banned in e.g. Denmark, it has in a spot check been detected on fruit and vegetables imported to Denmark Veterinary et al. (2003). The detection of the pesticide may be due to either contaminated soil, due to a long decomposition time, or due to legal/illegal use in some countries. The acaricide has commonly been applied to a number of fruits among others oranges, apples, grapefruit, lemon, mandarin, clementine, pears, table grape, exotic fruit and tomatoes Veterinary et al. (2003).

The goal of the third application to be discussed is hence a non-destructive, in-situ revelation of the banned pesticide, the organochlorine acaricide and insecticide dicofol, on tomato, apple and carrot by using a portable Raman spectrometer and SERS. A portable system will enable initial spot checking, e.g. in harbors or supermarkets.

As already mentioned, it is not unproblematic to apply SERS or SE(R)RS, since the molecular species should be in close proximity to a nano-structured metal surface in order to obtain an enhanced Raman signal and since the SERS or SE(R)RS spectra (the "molecular bar code") may be different from the Raman spectra of the free molecule. The Raman spectra can for some pesticides in a solution even with a concentration much higher than the detection limit not even be measured. If more pesticides are anticipated on a commodity, the more molecular information available and the more reliably the detection of even structurally related pesticides should be done. Besides, a quantitative analysis based on SERS/SE(R)RS requires in general special attention. Some of the approaches that can be used to ensure that the pesticide is brought reproducibly close to the surface are 1. the pesticide is attached to the molecular species functionalized to the surface of either the colloid or the substrate or 2. the molecules used to functionalize the surface provides molecular pockets, where the pesticide can be trapped close to the surface L. Guerrini (2008). The functionalization, may complicate the SERS spectra, which could necessitate the application of multivariate analysis to elucidate the SERS response from the molecular species under investigation (molecular bar code blurred due to interference from functionalization molecules).

The SERS spectra of an organochlorine pesticide deposited on an Ag substrate, similar to the one shown in figure 12, are earlier reported without any measures taken to ensure the molecular position close to the surface Alak & Vo-Dinh (n.d.). Applying the pesticide reproducibly to the surface itself is an act of art, especially when the foundation for a reproducible, quantitative measurement is required. This touches upon the question of how well the concentration is known, when the pesticide has been applied to the surface. In reference J. C. S. Costa & Corio (n.d.) the behavior of Au nano rods and Ag nano cubes as high performance SERS sensors has been evaluated for amongst others a chlorinated pesticide in a 10^{-7} M solution. In J. C. S. Costa & Corio (n.d.); L. Guerrini (2008) the SERS spectra were obtained by using research instruments.

In the present study a silver Film Over Nano spheres (AgFON) is used as the surface enhancing substrate and the portable Raman spectrometer (DeltaNu InspectorRaman shown in figure 2) as 'the analyzing unit'. The AgFON and periodic particle array (PPA) substrates fabrication process is shown in figure 12. This latter type of substrate can be fabricated according to the same principles governing the fabrication of AgFON's, the differences are, that the PPA's must be made on a transparent material, i.e. cover glass, and the final step is a lift-off of the nano spheres and the metal covering these, leaving only the triangular shaped deposits on the glass. Details about the fabrication process can be found in references A. Henry & Duyne (n.d.); Willets & Duyne (n.d.).

Conventional Raman spectra from the mentioned commodities were first acquired in order to be able to judge the the possible intervenience from the background spectra, see figure 13.

The normal Raman spectrum of the pure pesticide (figure 14) is initially recorded and compared to the pesticide reference spectrum obtained in reference M. L. Nicholas & Bromund (1976) and the obtained SERS spectra as well. In this way it is possible to detect a pesticide, which was not detected by using conventional Raman spectroscopy and to identify the individual peaks from the pesticide in the SERS spectrum.

Next, in figure 15 are seen a Raman spectrum of a 10^{-3} M pesticide dissolved in methanol and a SERS spectrum of the same solution as well as the spectrum of the bare nanostructured surface. The substrate of the type AgFON is seen in the figure as well.

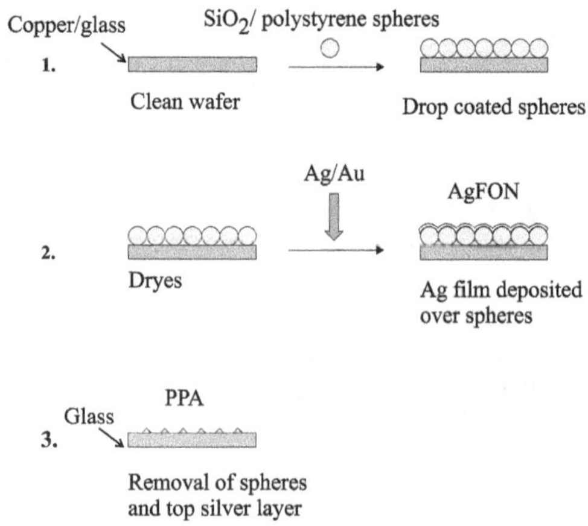

Fig. 12. A schematic representation of the fabrication procedure, when making AgFON's or PPA's using thermal vapor deposition (A. Henry & Duyne (n.d.); Willets & Duyne (n.d.)). In the case of the AgFON, the wafer used in the present research has been copper and the wafer used in the PPA is cover glasses. The metal used is Ag and the thickness of the layer 200 nm and the spheres used are polystyrene spheres.

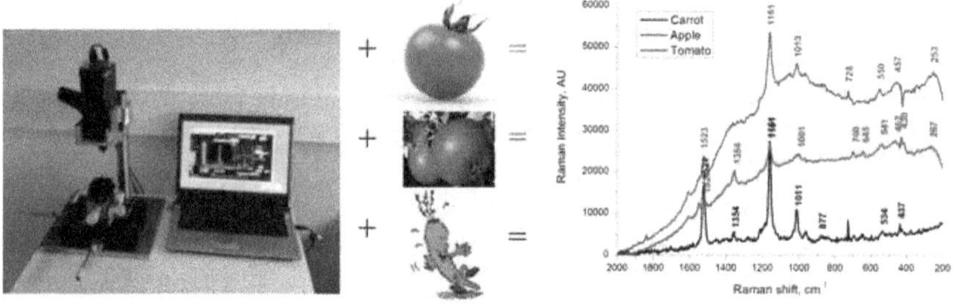

Fig. 13. Conventional Raman spectra of tomato, apple and carrot acquired with Inspector Raman.

The Raman spectrum of the pesticide solution reveals no presence of dicofol only the Raman spectrum of the solvent is observed underlining the need for enhancement. In the SERS spectrum the peak at 1096 cm^{-1} indicates the presence of dicofol, which is seen by comparing the spectrum of dicofol powder with the SERS spectrum. The SERS spectrum contain several Raman bands, but by comparing the spectrum of the bare substrate with the SERS spectrum, it is seen that these overlap with Raman bands found on the bare AgFON. This stresses the need for developing a handling/cleaning procedure or a protective layer for the substrates.

In the implementation of an on-site detection of pesticides on fruits and vegetables, the idea is that the substrate must be transmission based in stead of reflection based as used presently. This will possibly enable a closer contact with the fruit or vegetable surface and a penetration of the laser through the PPA to the commodity, the scattered Raman signal should be detected through the thin glass surface as well. Since dicofol, does not form covalent bonds with the surface, but interacts via other more weak forces, it is difficult to obtain

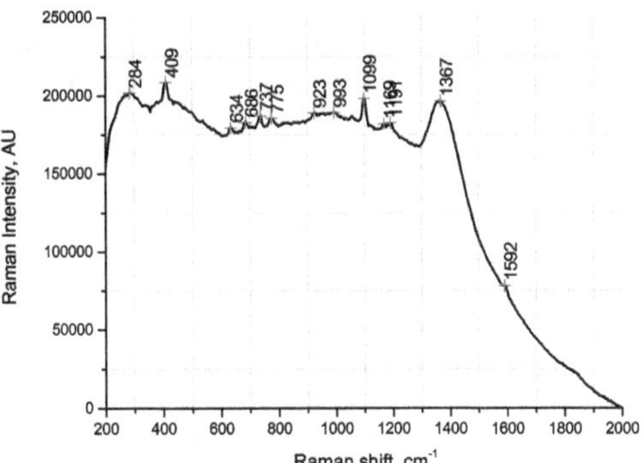

Fig. 14. A Raman spectrum of a pure dicofol powder, which was acquired with Inspector Raman. Notice the peak at 1099 cm^{-1}.

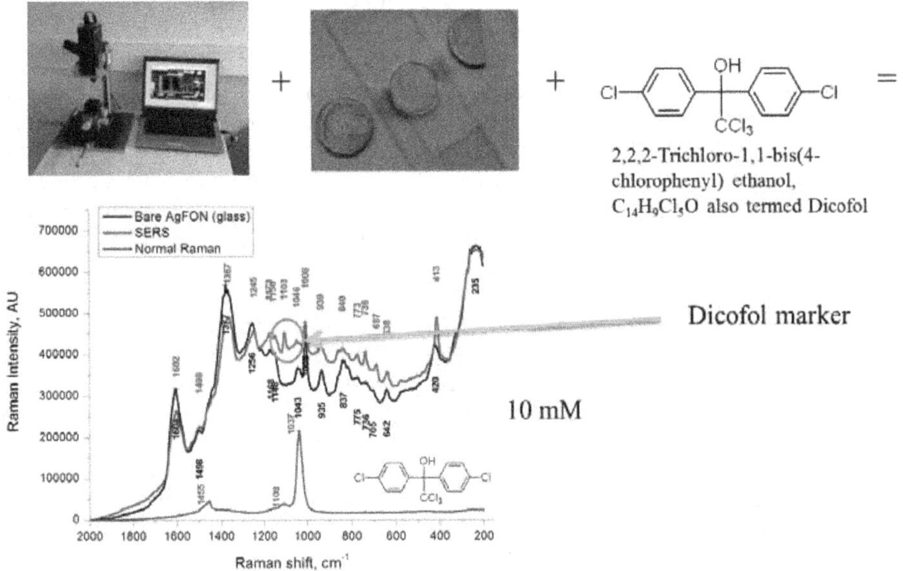

Fig. 15. Raman spectrum of a 10^{-3} M pesticide solution (blue), the SERS spectrum of the same solution (red) and a Raman spectrum of the bare AgFON. The solvent is methanol.

reliable and quantitative measurements without any measures taken to ensure a reproducible binding to the surface. A functionalization with albumin, exploiting the high lipid affinity of organochlorine pesticides, and/or albumin affinity, may be the answer to this problem M. Gülden & Seibert (2002); Maliwal & Guthrie (1981); Moss & Hathway (1964). Besides, a complete displacement of the surface contamination does not take place, which may cause a large background signal. The toxic nature of the pesticides complicates the optimization process of the SERS measurements on these.

In reference (n.d.b); C. Shende & S. Farquaharson (n.d.) it is demonstrated that single pesticides (phosphor- and nitrogen containing) and a mixture of pesticides can be detected fast (5 minutes) in low concentrations (50 - 100 ppm). The SERS setup included a portable FT-Raman spectrometer (785 nm) and specially prepared SERS active capillary probes for chemical extraction of the pesticides and generation of the SERS signal.

Further information about the experiments and fabrication of the substrates as well the possibilities for implementation can be obtained from the authors (kidje@bmb.sdu.dk).

4. Outlook

Raman spectroscopy has through many years been recognized as an advanced research tool for obtaining detailed molecular information. Depending on how the experiments are preformed different kind of information can be revealed such as molecular structure, dynamics and functions of bio-molecules . All due to the coherent nature of the Raman scattering process. Raman spectroscopy meets almost all the requirements listed in the introduction and has, due to the development of highly sensitive CCD-detectors, a variety of small laser sources, hand held spectrometers (dispersive and FT-instruments)and fast computers for the implementation of advanced data processing, the potential for being a fast, non-destructive and molecule specific tool for inspection of food quality. Despite this, Raman spectroscopy has not yet become a standard method within control of food quality, especially not as a standard on-site inspection technique. Our aim has not been to write an exhaustive review article about application of Raman scattering in food analysis Li-Chan (1996), but merely to present an adequate and pinpointed amount of theory and experimental aspects through which an increased understanding enables a more inspiring, creative and intelligent access to applying some kind of Raman spectroscopy within food applications. One major challenge is still to transform the Raman technique from research and analytical laboratories into real life applications, especially the detection of trace amounts of unknown, unwanted substances may be a challenge. We have demonstrated, through the discussion of three rather different case studies, that it is possible, but it requires additional developing work to be performed. Before one attempts to meet the practical challenges, it is therefore mandatory to understand the practical hurdles to be overcome in order to benefit from the high molecular selectivity that Raman spectroscopy can offer. Although there are remaining problems to be solved, we believe that within a few years Raman spectroscopy will develop into the area of analyzing food quality, on-site and non-destructively.

5. References

(n.d.a).

(n.d.b).

A. Henry, J. M. Bingham, E. R. L. D. M. G. C. S. & Duyne, R. P. V. (n.d.). *J. Phys. Chem. C* .

A. K. Smilde, R. B. & Geladi, P. (2004). *Multi-way Analysis: Applications in the Chemical Sciences.*

Alak, A. M. & Vo-Dinh, T. (n.d.). *Analytica Chimica Acta* .

C. Shende, F. Inscore, A. G. P. M. & S. Farquaharson, Journal = *Nondestructive Sensing for Food Safety, Quality, and Natural Resources, Proc. of SPIE 5587* Number = , P. . . T. . A. V. . .-I. . . M. . . Y. . . (n.d.).

Coultate, T. P. (2002). *food, the chemistry of its components*, number 4.

Danish Environmental Protection Agency, D. M. o. t. E. (2005). Bekæmpelsesmiddelstatistik 2005.

Danish Ministry of the Environment, E. P. A. (2005). Bekæmpelsesmiddelforskning fra miljøstyrelsen, kombinationseffekter af pesticider, 98.

eds. R. J. H. Clarke & Hester, R. H. (n.d.). *Advances in Infrared and Raman Spectroscopy (Vol. 1 - 12) and Advances in Spectroscopy (Vol. 13 and onwards)*, Vol. 1 - 12 and 13 and ongoing.

G. Jordan, R. Thomasius, H. S. J. S. W. O. S. B. S. M. M. H. S.-H. D. K. R. S. F. S. . K. D. L. (n.d.). *Journal für Verbraucherschutz und Lebensmittelsicherheit* .

H. D. Belitz, W. G. & Schieberle, P. (2004). *Food Chemistry*, number 3.

H. Schulz, M. B. & Baranski, R. (2005). Potential of nir-ft-raman spectroscopy in natural carotenoid analysis, *Biopolymers* (No. 77): 212 – 221.

Hedegaard, M. & Hassing, S. (2008). Application of raman dispersion spectroscopy in 3-way multivariate data analysis, *J. Raman Spectrosc.* Vol. 7(No. 39): 478 – 489.

Herrero, A. M. (n.d.). *Food Chemistry* .

J. C. S. Costa, R. A. Ando, A. C. S. L. M. R. P. S. S. M. L. A. T. & Corio, P. (n.d.). *Phys. Chem. Chem. Phys* .

Jernshøj, K. D. & Hassing, S. (2009). Analysis of reflectance and transmittance measurements on absorbing and scattering small samples using a modified integrating sphere setup, *Applied Spectroscopy* 8(63): 879–888.
URL: *http://www.opticsinfobase.org*

Jernshøj, K. D. & Hassing, S. (2010). Survival of molecular information under surfaced enhanced resonance raman (serrs) conditions, *J. Raman Spectrosc.* Vol. 7(No. 41): 727 – 738.

K. D. Jernshøj, S. Hassing, R. S. H. & Krohne-Nielsen, P. (2011). Experimental study on polarized se(r)rs of rhodamine 6g adsorbed on porous al_2o_3 substrates, *J. Chem. Phys.* (in print).

K. Davies, e. (2004). *Plant pigments and their manipulation*, Vol. 14.

K. Kneipp, M. M. & (Eds.), H. K. (2006). *Surface-Enhanced Raman Scattering, Physics and Application*.

Kortüm, G. (1969). *Reflectance Spectroscopy, Principles, Methods, Applications*.

L. Guerrini, A. E. Aliaga, J. C. J. S. G.-J. S. S.-C. M. M. C.-V. J. V. G.-R. (2008). Functionalization of ag nanoparticles with the bis-acridinium lucigenin as a chemical assembler in the detection of persistent organic pollutants by surface-enhanced raman scattering, *Analytica Chimica Acta* 624: 286 – 293.

Li-Chan, E. C. Y. (1996). The applications of raman spectroscopy in food science, *Trends in Food Science and Technology* 7: 361 – 370.

Long, D. A. (1969). *The Raman Effect, A Unified Treatment of the Theory of Raman Scattering by Molecules*.

Long, D. A. (2002). *Raman Spectroscopy*.

M. Gülden, S. Mörchel, S. T. & Seibert, H. (2002). Impact of protein binding on the availability and cytotoxic potency of organochlorine pesticides and chlorophenols in vitro, *Toxicology* 175: 201 – 213.

M. L. Nicholas, D. L. Powell, T. R. W. & Bromund, R. H. (1976). Reference raman spectra of ddt and five structurally related pesticides containing the norbornene group, *Journal of the AOAC* 59(1): 197 – 208.

Maliwal, B. P. & Guthrie, F. E. (1981). Interaction of insecticides with human serum albumin, *Molecular Pharmacology* 20: 138–144.

Martin Hedegaard, Christoph Krafft, H. J. D. L. E. J.-S. H. & Popp, J. (2010). Discriminating isogenic cancer cells and identifying altered unsaturated fatty acid content as associated with metastasis status, using k-means clustering and partial least squares-discriminant analysis of raman maps, *Analytical Chemistry* 82(7): 2797 – 2802.

McCreery, R. L. (2000). *Raman Spectroscopy for Chemical Analysis.*

Miller, E. S. (1934). Quantative absorption spectra of the common carotenoids, *Plant Physiol.* 9(3): 693Ű694.

Mortensen, O. S. (1981). Raman dispersion spectroscopy (radis), 1 - phenomenology, *J. Raman Spectroscopy* 11(5): 321 – 333.

Mortensen, O. S. & Hassing, S. (1980). *Polarization and Interference Phenomena in Resonance Raman Scattering in Advances in Infrared and Raman Spectroscopy, R. J. H. Clark and R. E. Hester (eds.)*, Vol. 6.

Moss, J. A. & Hathway, D. E. (1964). Transport of organic compounds in the mammal, *Biochem. J.* 91: 384.

Plazek, G. (1934). *Rayleigh-Streuung und Raman Effekt, in Handbuch der Radiologie, E. Marx (ed.)*, Vol. 3048.

R. K. Dutta, P. K. S. & Pandey, A. C. (2009). Surface enhanced raman spectra of escherichia coli cells using zno nanoparticles, *Digest Journal of Nanomaterials and Biostructures* 4(1): 83 – 87.

R. L. Eriksen, A. Pors, J. D. A. C. S. & Albrektsen, O. (2010). Fabrication of large area homogenous metallic nanostructures for optical sensing using colloidal lithography, *Microelectronic Engineering* 87: 333 – 337.

Raman, C. V. & Krishnan, K. S. (1928). A new type of secondary radiation, *Nature* 3048(121): 501.

Ru, E. C. L. & Etchegoin, P. G. (2009). *Principles of Surface-Enhanced Raman Spectroscopy and related plasmonic effects.*

S. Hassing, K. D. Jernshøj, M. H. (2011). Solving chemical classification problems using polarized raman data, *J. Raman Spectrosc.* 42(1): 21 – 35.
URL: *onlinelibrary.wiley.com/doi/10.1002/jrs.2666/pdf/*

Smith, E. & Dent, G. (2005). *Modern Raman Spectroscopy, A Practical Approach.*

Veterinary, D. & Administration, F. (2003). Pesticidrester i foedevarer, bilag 2, analysemetoder anvendt i undersøgelser 2003.

Veterinary, D., Food Administration, M. o. F. & Affairs, C. (2003). Pesticides, food monitoring, 1998-2003, part 2.

Veterinary, D., Food Administration, M. o. F. & Affairs, C. (2005). Pesticidrester i fødevarer **2005** - resultater fra den danske pesticidkontrol.

Vidi, S. (2003). *Fourier-transform spectroscopy instrumentation engineering.*

W. R. Premasiri, D. T. Moir, M. S. K. N. K. G. J. & Ziegler, L. D. (2005). Characterization of the surface enhanced raman scattering (sers) of bacteria, *J. Phys. Chem. B* 109: 321 – 320.

Willets, K. A. & Duyne, R. P. V. (n.d.). *Ann. Rev. Phys. Chem.* .

Zeiri, L. (2007). Sers of plant material, *J. Raman Spectrosc.* 38: 950 – 955.

5

Contamination of Foods by Migration of Some Elements from Plastics Packaging

O. Al-Dayel, O. Al-Horayess, J. Hefni, A. Al-Durahim and T. Alajyan
King Abdulaziz City for Science and Technology, Riyadh, Saudi Arabia

1. Introduction

There are various types of packaging material including paper, board, plastic, metal, glass, wood and other materials. Paper and board packaging accounted for the largest share of global packaging sales in 2003 with 39% of the total. Plastic packaging accounted for 30% (rigid and flexible plastics) of the market, with metal packaging accounting for 18% and glass packaging a further 7%. Other packaging products accounted for the remaining 6% of the market. Rigid plastics was the fastest growing sector of the market during the period 1999-2003. Around 70% of overall consumer packaging consumption is used for food and beverage packaging (WPO, 2008). Many different types of plastics are being used as packaging materials. The key components in plastic materials are polymers which are made of units of organic material, and one or more of large molecular weight can be formed as desired. Most polymers are petrochemical compounds with additive materials to give them properties of flexibility, elasticity and resistance to fracture and transparency to light (Oi-Wah & Siu-Kay, 2000; Al-Dayel et. al. 2009)

The final plastic material thus is a mix of polymer, additives, manufacturing aids, and side products from the complex polymerization process that were not intentionally added (Bradley & Coulier 2007).

Low density Polyethylene (LDPE) is used in preparation of most of the hot food packaging. LDPE has high flexibility, and can be affected by organic solvents. It has melting temperature of 110 ° C (wikipedia.org 2012).

High Density Polyethylene (HDPE), has the same uses as those of the low density, but it is much flexible and resistant to organic solvents and to high temperatures. It is used in manufacture of some household appliances, pipes and hoses. It is also used in food packaging which are subject to sterilization temperatures. HDPE is characterized by its ability to isolate the humidity, and its flexibility even at freezing temperature.

Different types of additives, such as antioxidants, stabilizers, lubricants, anti-static and anti-blocking agents, have been developed to improve the performance of polymeric packaging materials (Achilias 2007; Susan1992).

The role of food and beverage packaging as a source of contaminants have raised many concerns after the widespread use of such containers; packaging. Any substance (monomers and other starting substances, additives, residues) which migrates from the packaging into the food is of concern if it could be harmful to the health (Donatella et. al., 2010; Grob et. al., 2006).

The migration of additives or contaminants from polymeric food packaging to food may be separated into three different, but inter-related, stages: diffusion within the polymer, solvation at the polymer food interface, and dispersion into bulk food (Oi-Wah & Siu-Kay, 2000). The migration has been showed to increase with fat content and storage temperature (Sanches, et al.,2007).

Antimony does not bioaccumulation, so exposure to naturally occurring antimony through food is very low. Antimony is present in food, including vegetables grown on antimony-contaminated soils, mostly in the low μg/kg wet weight range or less (WHO, 2003).

Antimony toxicity is dependent on the exposure dose, duration, route (breathing, eating, drinking, or skin contact), other chemical exposures, age, sex, nutritional status, family traits, life style, and state of health (Ross and Adrian, 2009). Chronic exposure to antimony in the air at levels of 9 mg/m^3 may exacerbate irritation of the eyes, skin, and lungs (Roper, 1992). Long-term inhalation of antimony can potentiate pneumoconiosis, altered electrocardiograms, stomach pain, diarrhea, vomiting, and stomach ulcers, results which were confirmed in laboratory animals (Roper, 1992). Although there were investigations of the effect of antimony in sudden infant death syndrome, current findings suggest no link. Long-term exposure in experimental animals has shown an increase in the hepatic malfunction and blood changes (ATSDR, 1992). It is not clear yet whether antimony is a human carcinogen. Occupational epidemiology could not confirm evidence of lung carcinogenicity caused by antimony as detected in female rats (Gerhardsson et al. 1982; Jones, 1994; Groth et. al.1986). Furthermore, because the experimental results were not uniform, animal lung carcinogenicity by antimony is still a matter for debate (Jones, 1994; Newton et. Al., 1994; Ross and Adrian, 2009).

This work is an attempt to evaluate the migration of elements from packaging materials into food stuffs. A severe contact condition between food and packaging materials has been created to evaluate the maximum potential migration of elements (Grob et. al.1999).

2. Experimental

2.1 Samples

Two types of polyethylene samples have been selected from the Saudi market. One is called thermal bag which is commonly used for hot food, and the other one is called food bag which is commonly used for cold and freezing food. Both of these types are made of liner low density polyethylene. The samples elemental concentrations have been obtained using neutron activation analysis (NAA).

2.2 Neutron Activation Analysis (NAA)

Samples were irradiated at the McMaster University Reactor in Hamilton, Ontario, Canada. It is a 2MW open pool research reactor.

The Gamma ray spectrum acquisition was carried out by the use of a high resolution intrinsic germanium detector.

Two portions of each sample were weighed into a plastic irradiation container. One portion is for short-lived isotope analysis (half life < 24 hours). This portion is about 4 grams weighed into a 7 ml volume container, and it was irradiated for about 60 seconds using a thermal neutron flux of approximately $6x10^{12}$ n/ cm^2.s . After a short decay period of 6 minutes, the gamma ray spectrum was acquired. The sample was then allowed to decay for further period of 24 hours.

The other portion was for longer lived isotope analysis (half life >5 days). Approximately 24 grams of the sample was weighed into a 40 ml container, and was irradiated for 20 minutes using thermal neutron flux of approximately $8x10^{12}$ n/ cm^2.s. After 3 days of decay, the sample was counted for 60 minutes. The sample was again allowed to decay for further period of 18 days and then counted for 2 hours. Results were calculated for each of the four spectra.

2.3 Quality assurance for NAA analysis

To assess the analytical process and make a comparative analysis, Standard Reference Material coal sample (SARM-18) from South African Bureau of Standards and a trace element in coal sample from the USA National Institute of Standards and Technology (NIST) (SRM 1632c) were analyzed in the same manner as other samples (Anderson & Cunningham, 2000; Wang & Sakanishi, 2004). Table (1) gives the comparison of the certified values and these obtained in this work for each reference materials. The results are generally in a good agreement except for Sr, Zr, Ba, Sb, Ba, Ca, Eu, Na, and As in one or other standard reference material samples.

Element in mg/kg (%) (µg/Kg)	SARM 18			NIST 1632c		
	This work	RSD	Range@	This work	RSD	Certified Value
Al (%)	1.34	3	*2.54-2.61*	0.94	4	0.915±0.0137
Sb	0.30	5	*0.3**	0.22	5	0.461±0.029
As	0.52	4		3.5	3	6.18±0.27
Ba	91	6	*71-82*	70	5	41.1±1.6
Br	3.5	3	*3**	20.9	3	18.7±0.4
Cd	0.7	20		<0.3		0.072±0.007
Ca (%)	0.138	4	*0.17-0.19*	0.21	4	0.145±0.03
Ce	20.6	3	*21-24*	8.2	3	11.9±0.2
Cs	1.17	4	*1**	0.473	5	0.594±0.010
Cl	54	5		1130	3	
Cr	17.3	3	*14-18*	10.4	3	13.73±0.20
Co	7.5	3	*5.5-7.2*	3.4	3	3.48±0.2
Cu	4.9	9		6.7	6	6.01±0.25
Dy	1.87	3		0.57	3	

Eu	0.31	3	0.3*	0.17	3	0.124±0.003
Ga	7.1	5	8*	2.6	6	3*
Au (µg/Kg)	0.65	23		<0.5		
Hf	1.9	3	1.7-1.9	0.44	4	0.585±0.010
In	0.028	12		0.02	11	
I	1.0	11		1.5	6	
Ir (µg/Kg)	<0.7			<0.5		
Fe (%)	0.22	3	0.28-0.29	0.77	3	0.735±0.011
La	10.6	3	9.0-13	4.7	3	
Lu	0.207	3		0.055	4	
Mg (%)	0.056	6.5	0.1 0.11	0.039	8	0.0384±0.0032
Mn	21.5	3.2	21-23	13.5	3	13.04±0.53
Hg	<0.017		0.04*	0.0802	8	0.0938±0.0037
Mo	1.13	5		0.693	9	0.8*
Nd	7.5	5		3.1	5	
Ni	<10			<6.8		9.32±0.51
K (%)	0.12	4	0.140-0.150	0.081	3	0.11±0.0041
Rb	8.3	6	6.7-9.5	5.4	6	7.52±0.33
Sm	1.7	5	1.9-2.2	0.72	5	1.078±0.028
Sc	5.1	3	4.0-4.7	2.0	3	2.905±0.036
Se	0.2	31		1.1	5	1.326±0.071
Ag	<0.2			<0.1		
Na (%)	0.013	3		0.052	3	0.03±0.005
Sr	31	15	42-45	109	4	63.8±1.4
Ta	0.36	3	0.3*	0.14	3	
Te	<0.9			<0.3		0.05*
Tb	0.29	6	0.3*	0.10	9	
Th	3.8	3	3.0-4.3	1.2	3	
Sn	<11		1*	<7		1*
Ti	0.067	3	0.111-0.116	0.047	4	0.052±0.003
W	1.5	3	2*	0.4	6	
U	1.9	3	1.5-2.0	0.40	4	0.513±0.012
V	22	3	21-25	15	4	23.72±0.51
Yb	1.26	3		0.34	4	
Zn	6	10		10	6	12.1±1.3
Zr	20	19	62-71	<15		16*

@ Range refers to 95% confidence limits.
* Uncertified value.
RSD: Relative Standard Deviation.

Table 1. Comparison of elemental concentration of the standard reference material (SARM18 and NIST 1632c) in this work and their certified values.

2.4 Study of elemental migration

Four types of food stuffs were examined, water, 30% ethanol, olive oil and 5% acetic acid. The aim was to study the immigration of Al, Sb, Cu, Mg, Ti and Zn.

To assess the maximum possible value of elements migration, the samples were prepared under the influence of severe contact conditions between the bag materials and the food stuff. This influence have been created by mixing the bag material, in the form of powder, with the food stuffs and exposing the mixture to high temperature (80-100 °C). Such a contact case cannot be exist in normal human use, but it can show the maximum potential migration.

2.5 Experimental arrangement and ICP-MS analysis

Four samples were prepared in four flasks. Each flask contained a mix of 2.5 g of the sample powder and 25 ml of one of the four food stuffs. The flasks were placed in a shaker at 160 rpm stirring rate at temperature of 100 °C (80 °C for ethanol case).

The samples were elementally analyzed using a Perkin-Elmer Sciex Instruments multi-element ICP-MS spectrometer, type ELAN6100, equipped with a standard torch, cross flow nebulizer and Ni sampler and skimmer cones.

2.6 Quality assurance for ICP-MS analysis

To assess of the analytical process and make a comparative analysis, Standard Reference Material (SRM), Nist-1640 Natural Water purchased from the National Institute of Standards and Technology (NIST), USA was analyzed in the same manner as all other samples. Table 2 compare the certified values with those obtained in this work. The results are generally in good agreement with the certified values.

3. Results and discussion

The elemental concentrations in the samples and the corresponding concentrations migrated to food stuffs are shown in table 3. The results show that the Antimony (Sb) has the highest migrated concentration ratio (37 %), in the acetic acid, followed by Zink (Zn) with 22 % , Magnesium with 17 %, in ethanol and Titanium (Ti) with 12%, in olive oil. The lowest migration ratio was that of Aluminum (Al) which was only 0.012 %, in ethanol. Table 3: Concentration of elements in bags materials and migrated to food stuffs.

4. Conclusion

The NAA and the ICP-MS analytical methods used in this work gives good results. These results were confirmed by the analysis of the standard reference materials as shown in tables 1 and 2.

Migration of substances from plastic packaging materials into food stuffs is clearly measured in the conditions which have been created to represent the worst case of contact between packaging material and food stuff. This suggests that a further studies on migration of substances from plastic packaging into food stuffs in the normal conditions should tack place. The results also suggest expected a seriousness harmful effect of using wrong plastic packaging material for heating foods in microwave ovens.

Elements	This Work			Certified values	
	Concentration in ppb	RSD		Concentration in ppb	RSD
Li	49.8	1.50		50.7	2.76
Be	36.4	4.95		34.94	1.17
B	264	8.60		301.1	2.03
Na	27800	0.34		29530	1.05
Mg	5800	0.50		5819	0.96
Al	47.4	0.32		52	2.88
K	899	0.93		994	2.72
Ca	6990	0.97		7045	1.26
V	11.9	1.16		12.99	2.85
Cr	37.1	0.45		38.6	4.15
Mn	117	0.46		121.5	0.91
Fe	35.2	15.65		34.3	4.66
Co	19	0.50		20.28	1.53
Ni	27	1.19		27.4	2.92
Cu	89.2	0.19		85.2	1.41
Zn	64.7	1.36		53.2	2.08
Ga	0.198	30.20			
As	27.3	1.59		26.67	1.54
Se	23.7	3.95		21.96	2.32
Rb	2.22	1.65		2	1
Sr	108	0.40		124.2	0.56
Mo	43.5	2.04		46.75	0.56
Ag	6.48	0.39		7.62	3.28
Cd	21.9	2.14		22.79	4.21
Sb	12.6	2.85		13.79	1.46
Te	0.328	26.89			
Ba	139	0.96		148	1.48
Tl	0.146	42.60		<0.1*	
Pb	26.8	0.79		27.89	0.50
Bi	0.143	40.49			
U	0.834	7.41			

RSD: Relative Standard Deviation.

Table 2. Comparison of elemental concentration of the reference material (NIST 1640) in this work with the certified values.

Element	Concentration of elements in bags materials	concentation of element migrated to food material and the migration percentage							
	ppm	ppm	%	ppm	%	Ppm	%	ppm	%
Thermal bag		Water	Migration	Ethanol	Migration	Olive Oil	Migration	Acetic Acid	Migration
Al	100	0.0147	0.15	0.0127	0.13	0.073	0.7	0.181	1.8
Sb	0.362	0.00307	8.48	0.00372	10.28	0.000631	1.7	0.00824	22.8
Cu	2.7	0.00668	2.47	0.0076	2.81	0.00224	0.8	Not Detect	
Mg	246	0.0461	0.19	0.553	2.25	0.014	0.1	Not Detect	
Ti	6.6	0.0101	1.53	Below Detection Limit		0.00665	1.0	Below Detection Limit	
Zn	98	0.0175	0.18	0.0652	0.67	0.0206	0.2	0.982	10.0
Food bag									
Al	768	0.0968	0.13	0.0247	0.03	0.244	0.3	0.413	0.5
Sb	0.283	0.00372	13.14	0.00391	13.82	0.00066	2.3	0.0105	37.1
Cu	4.75	0.00638	1.34	0.0154	3.24	0.00661	1.4	Not Detect	
Mg	32	0.0274	0.86	0.555	17.34	0.0374	1.2	Not Detect	
Ti	4.4	0.0115	2.61			0.055	12.5	0.000447	0.1
Zn	52	0.026	0.50	0.124	2.38	0.163	3.1	1.15	22.1

Table 3. Concentration of elements in bags materials and migrated to food stuffs.

5. References

Achilias D. S., Roupakias C., Megalokonomosa P., Lappas A.A., Antonakou E.V. (2007) Chemical recycling of plastic wastes made from polyethylene (LDPE and HDPE) and polypropylene (PP) ElsevierB.V. (available online)

Agency for Toxic Substances and Disease Registry (ATSDR), (1992) *Toxicological profile for antimony.* Atlanta, Georgia, USA: US Department of Health and Human Services, Public Health Service; 1992. p. 160

Al-Dayel O., Al-Horayess O., Hefni J. and Al-Durahim A.,(2009) Trace Elements in Packaging Polymers, Research Journal of Chemistry and Environment, 13, 1, 92

Anderson D and Cunningham W, (2000) Revalidation and Long-Term Standard Reference Materials, J.AOAC.Int.,83, 5, 1121

Bradley, E and Coulier, L, (2007) An investigation into the reaction and breakdown products from starting substances used to produce food contact plastics, Central Science Laboratory, London.

Donatella Restuccia , U. Gianfranco Spizzirri , Ortensia I. Parisi, Giuseppe Cirillo, Manuela Curcio, Francesca Iemma, Francesco Puoci, Giuliana Vinci, Nevio Picci, (2010) New EU regulation aspects and global market of active and intelligent packaging for food industry applications, *Food Control* 21, 1425

Gerhardsson L, Brune D, Nordberg G.F, Wester P.O, (1982) *Antimony in lung, liver and kidney tissue from deceased smelter workers, Scand. J. Work Environ. Health*, 8 , pp. 201–208

Grob K, Spinner C, Brunner M, Etter R. (1999) The migration from the internal coatings of food cans; summary of the findings and call for more effective regulation of polymers in contact with foods: a review. Food Addit Contam;16. 579

Grob K, Biedermann M, Scherbaum E, Roth M, Rieger K. (2006) Food contamination with organic materials in perspective: packaging materials as the largest and least controlled source? A view focusing on the European situation. Crit Rev Food Sci Nutr;46, 529

Groth D. H, Stettler L.E , Burg J.R, Busey W.M, Grant G.C, Wong L,(1986) *Carcinogenic effects of antimony trioxide and antimony ore concentrate in rats*, J. Toxicol. Environ. Health, 18), pp. 607–626

Jones R.D , (1994) Survey of antimony workers: mortality 1961–1992, Occup. Environ. Med., 51 , pp. 772–776

Newton P.E., Bolte H.F., Daly I.W., Pillsbury B.D., Terrill J.B., Drew R.T., Ben Dyke R, Sheldon A.W., Rubin L.F.,(1994) *Subchronic and chronic inhalation toxicity of antimony trioxide in the rat*, Fundam. Appl. Toxicol., 22, 561–576

Qi-Wah Lau , Siu-Kay Wong (2000) , Contamination in food from packaging material, *Journal of Chromatography* A, 882, 255

Roper WL, editor. (1992) *Toxicological Profile for Antimony and Compounds.* Atlanta, Georgia, USA: Agency for Toxic Substances and Disease Registry. Public Health Statement; pp. 1–5

Ross G. Cooper and Adrian P. Harrison, (2009 April) *The exposure to and health effects of antimony*, Indian J Occup Environ Med. 13(1): 3–10

Sanches Silva, A, *et al.*,(2007) Kinetic migration studies from packaging films into meat products, *Meat Science* 77, 238

Susan Brewer M. (1992). , Cooperative Extension Service, University of Illinois at Urbana-Champaign, Circular 1320A.

Wang J, Nakazato T and Sakanishi K, (2004) Microwave digestion with HNO3/H2O2 mixture at high temperatures for determination of trace elements in coal by ICP-OES and ICP-MS, J. Analytical Chimica Acta, 514, 115

WHO/SDE/WSH/03.04/74, (2003) Antimony in Drinking-water

WPO (2008) Market Statistics and Future Trends in Global Packaging

WPO – World Packaging Organisation / PIRA International Ltda.

www.wikipedia.org (2012)

Section 2

Some Case Studies Improving the Food Quality

6

Senescence of the *Lentinula edodes* Fruiting Body After Harvesting

Yuichi Sakamoto[1], Keiko Nakade[1,2],
Naotake Konno[1] and Toshitsugu Sato[1,3]
[1]*Iwate Biotechnology Research Center,*
[2]*TSUMURA & CO*
[3]*Kitami Institute of Technology,*
Japan

1. Introduction

Lentinula edodes, or shiitake mushroom as it is more popularly known, is one of the most economically important edible cultivated mushrooms. However, postharvest spoilage, such as browning of the gills and softening of the fruiting body, results in loss of freshness and consequent loss of food value (Minamide et al.; 1980a, b).

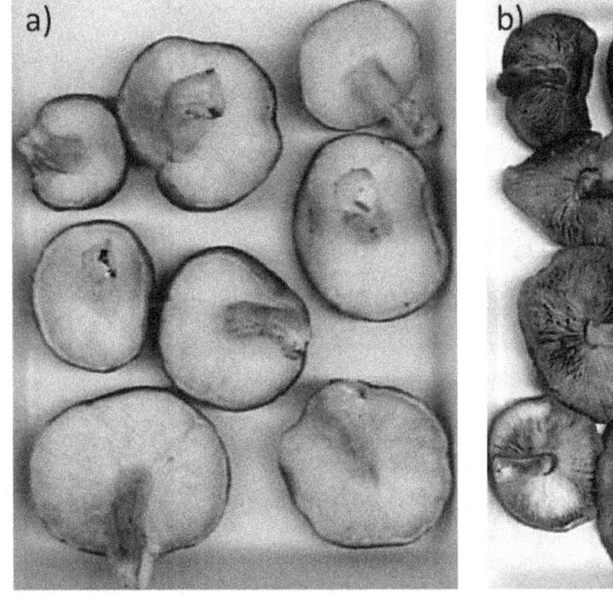

Fig. 1. Fruiting bodies a): just after harvest (fresh fruiting body) and b): at 4 days after harvest (senescent fruiting body).

Numerous studies addressing the mechanisms of quality loss during postharvest storage have revealed that browning of the *L. edodes* fruiting body is associated with increased activities of tyrosinase (Tyr; Kanda et al., 1996a, b) and laccase (Lcc; Nagai et al., 2003) after harvest. In addition to being an important fresh food source, shiitake mushrooms also have medicinal value. Lentinan, a β-1,3-glucan used for tumor immunotherapy (Chihara et al., 1969), is purified from fresh shiitake mushrooms. However, lentinan content decreases during postharvest storage (Minato et al., 1999). The reported structure of lentinan is a β-1,3-linked-D-glucan with β-1,6 branches (Chihara et al., 1969), and it appears that postharvest degradation of lentinan during mushroom storage is mediated by β-1,3-glucanase (Minato et al., 1999). Four glucanases have been reported in *L. edodes* fruiting bodies, two exo-β-1,3-glucanases, EXG1 (Sakamoto et al., 2005a) and EXG2 (Sakamoto et al., 2005b), and two endo-β-1,3-glucanases, TLG1 (Sakamoto et al., 2006) and GLU1 (Sakamoto et al., 2011). An endo-β-1,6-glucanase, Pus30, was also purified from *L. edodes* fruiting body (Konno and Sakamoto, 2011). Except for EXG1, these glucanases are involved in lentinan degradation after harvesting.

Postharvest changes are considered to be complex, and there is currently little information about the changes in gene transcription following harvest of the *L. edodes* fruiting bodies. It has been reported that the expression of several genes increases during the postharvest period in *Agaricus bisporus* (button mushroom, Eastwood, 2001). Eastwood et al., (2001) revealed that argininosuccinate lyase increases after harvest, and the relationship between elevated levels of argininosuccinate lyase and postharvest physiological changes of the *A. bisporus* fruiting body has been investigated in detail (Wagemaker et al., 2007; Eastwood et al., 2008). Several studies have analyzed changes in gene transcription during development of the fruiting body of *L. edodes* (Hirano et al., 2004; Miyazaki et al., 2005). More recently, EST analysis (Suizu et al., 2008) and SAGE analysis (Chum et al., 2008, 2011) were carried out on *L. edodes*. Changes in gene transcription after harvest of the *L. edodes* fruiting body were investigated, revealing that many genes were newly expressed after harvest, such as putative chitinases, chitosanase, and a transcription factor (Sakamoto et al., 2009). This chapter will discuss about genes involved in fruiting body senescence in mushrooms, especially in *L. edodes*.

2. Phenol oxidases involved in browning of fruiting body after harvesting

The postharvest preservation of *L. edodes* fruiting bodies causes gill browning, which is commercially undesirable since it causes an unpleasant appearance. In general, melanins involved in gill browning are considered to be synthesized from β-(3,4-dihydroxyphenyl)alanine (DOPA), derived from tyrosine. DOPA can be oxidized enzymatically to quinones, which polymerize nonenzymatically to form the melanin pigments. Oxidation of tyrosine is commonly catalyzed by tyrosinasae (Tyr: EC1.14.18.1). The mechanisms of mushroom browning have been investigated extensively in *A. bisporus* (Burton, 1988; Espín et al., 1999). Browning in this species is mainly due to DOPA and c-glutaminyl-3,4-dihydroxy- benzene (GDHB) melanins (Jolivet et al., 1998), and Tyr seems to play the most important role in their synthesis (Turner, 1974). Burton (1988) reported that epidermal tissues of *A. bisporus* had a greater activity of non-latent Tyr and a greater concentration of phenols than did the flesh. It has been reported that Tyr activity of *L. edodes* fruiting bodies increases after harvest, and a Tyr has been purified (Kanda et al., 1996a).

Laccases (Lccs: EC 1.10.3.2) catalyze the single-electron oxidation of phenols or aromatic amines to form different products via various pathways. Lcc belongs to a group of polyphenol oxidases that contain copper atoms in their catalytic center; thus, they are typically referred to as multicopper oxidases. Lcc catalyzes the single-electron oxidation of phenolic substrates or aromatic amines to form different products via a variety of biochemical pathways (Leonowicz et al., 2001). There are many reports in the literature of the purification and characterization of Lcc isoforms from white-rot fungi, and several Lcc-encoding genes have been isolated (reviewed by Kumar et al., 2003; Baldrian, 2006). A correlation between melanin synthesis and intracellular Lcc in *Cryptococcus neoformans* has been reported (Ikeda et al., 2002). Lcc in *Aspergillus nidulans* is considered to form L-DOPA to synthesize melanins in conidia (Aramayo and Timberlake, 1990). Activity of Lcc increases after harvest of the *L. edodes* fruiting body, and a Lcc in *L. edodes* purified from fruiting bodies after harvest can oxidize DOPA (Nagai et al., 2003). Therefore, it is considered that Tyr and Lcc are involved in melanin synthesis after harvest of *L. edodes* fruiting bodies.

2.1 Tyr involvement in gill browning after harvest

Tyr (EC 1.14.18.1) is an ubiquitous enzyme in nature and the key enzyme in the process of melanin biosynthesis (van Gelder et al., 1997). Tyr catalyzes oxidation of phenolic substrates to quinone, which spontaneously polymerizes into dark-colored pigments known as melanin in the presence of nucleophilic moieties. Tyrs are found in a wide range of organisms, including prokaryotic and eukaryotic microorganisms, plants, invertebrates, and mammals. Tyrs are involved in a variety of biological functions, for example skin pigmentation in mammals and browning in plants and mushrooms. In mushrooms, browning after harvest has been investigated, especially in *A. bisporus* and *L. edodes*. Browning after harvest is considered a consequence of the Tyr-catalyzed oxidation of phenolic substrates (e.g., DOPA) into quinones leading to the formation of dark pigments of melanins. The enzymatic pigmentation of mushrooms is mediated largely by Tyr (Jolivet et al., 1998).

Several reports revealed that Tyr is related to gill browning after harvest of *L. edodes* fruiting bodies. It was reported that Tyr activity increases in the gills during postharvest preservation (Kanda et al., 1996a). A Tyr (LeTyr) was purified and characterized as the *L. edodes* Tyr, and LeTyr can catalyze tyrosine to DOPA (Kanda et al. 1996a, 1996b). The gene encoding LeTyr (*Letyr*) was isolated, and anti-serum was synthesized to LeTyr (Sato et al., 2009). Sato et al., (2009) showed that LeTyr increased after harvest, suggesting that LeTyr catalyzes DOPA synthesis for melanin formation after harvest in *L. edodes*. *Letyr* is the only Tyr encoding gene in *L. edodes* (Sato et al., 2009). On the other hand, Tyr is involved in browning of the surface of vegetative mycelia (Sano et al., 2010), suggesting that LeTyr is involved in melanin synthesis in different tissues. Browning of the surface of vegetative mycelia is regulated by light, and the blue light receptor PHRB regulates *Letyr* (Sano et al., 2009). This suggests that melanin synthesis in vegetative mycelium under light is caused by an increase of LeTyr via PHRB. PHRB is a homolog of WC2 in *Neurospora crassa,* and expression of the WC2 homolog in *L. edodes* is low in the *L. edodes* fruiting body after harvest (Sakamoto et al., 2009). This suggests that expression of *Letyr* in the fruiting body after harvest is regulated by transcription factor(s) other than PHRB.

2.2 Lcc involvement in gill browning after harvest

White-rot fungi produce several isoforms of extracellular lignin degrading enzymes, including lignin peroxidase, manganese peroxidase, and Lcc. These lignin degrading enzymes are considered as secreted enzymes, but several Lccs are intracellular enzymes.

Two Lccs (Lcc1 and Lcc2) have been purified from *L. edodes* (Nagai et al., 2002 and 2003), an Lcc-encoding gene has been cloned and characterized, and six Lcc encoding genes (*lcc1-lcc6;* lcc1: AB035409; lcc2: AB035410; lcc3: AB046713; Lcc4: AB446445; Lcc5: AB543788; Lcc6: AB543787) have been deposited in the DNA data bank (DDBJ) (Zhao and Kwan, 1999; Sakamoto et al., 2008, 2009; Yano et al., 2010). Lcc1 is secreted from vegetative mycelia in culture (Nagai et al., 2002) and is encoded by *lcc1* (Sakamoto et al., 2008). Lcc2 is expressed in the brown gills of fruiting bodies after harvesting (Nagai et al., 2003). The putative amino acid sequence of *lcc4* includes identical amino acid sequences to the N-terminal amino acid sequence of an enzymatically digested peptide of Lcc2 (Sakamoto et al., 2009). We expressed the *lcc4* gene heterologously in *Aspergillus oryzae*, and observed Lcc activity of the recombinant enzyme (Yano et al. 2009). These data suggest that *lcc4* encodes Lcc2, as designated by Nagai et al. (2003), and the gene *lcc2* (Accession No. AB035410) does not. Therefore, the Lcc2 purified by Nagai et al. (2003) is designated Lcc4 in this paper. Expression of *lcc1* is high in vegetative mycelia (Sakamoto et al., 2008), but *lcc4* is expressed only in the fruiting body and increases after harvest (Sakamoto et al., 2009). Furthermore, Lcc1 cannot oxidize DOPA (Nagai et al., 2002), but Lcc4 can (Nagai et al., 2003; Yano et al., 2009). These observations collectively suggest that Lcc4 is involved in melanin synthesis after harvesting by catalyzing L-DOPA to DOPA quinone. Gill browning after harvesting of *L. edodes* fruiting bodies is considered to be caused by melanin synthesis due to cooperation of LeTyr and Lcc4 and their increased levels after harvest.

3. Cell wall degrading enzymes involved in fruiting body autolysis

Fruiting body softening occurs due to cell wall degradation. The cell wall of *L. edodes* is constructed of several polysaccharides, such as β-1,3-glucan, β-1,6-glucan, chitin, and chitosan (Shida et al. 1981). Thus, β-1,3-glucanase, β-1,6-glucanase, chitinase, chitosanase are involved in cell wall degradation after harvesting. Cell wall degrading enzymes found in the *L. edodes* fruiting body after harvest will be introduced. One of the cell wall components, β-1,3-1,6-glucan, called lentinan, is used for antitumor therapy. Lentinan degradation after harvesting is caused by an increase in β-1,3-glucanase activity after harvesting. Thus, controlling β-1,3-glucanase expression is very important to keep the lentinan content in the *L. edodes* fruiting body after harvest. There are few reports on cell wall degrading enzymes, such as endoglucanase and exoglucanase encoding in basidiomycetous fungi, such as in *A. bisporus*. However, there were fewer reports on glucanases related to senescence of mushrooms when we started our research on senescence in *L. edodes*. There were several studies on chitinase activity in several mushrooms, but there was no evidence of a relationship between chitinases and mushroom senescence at that time. Cell wall degrading enzymes in *L. edodes* fruiting bodies after are discussed in this section.

3.1 β-glucanases

The cell wall structure of fungi is constantly changing during mycelial growth and the cell cycle. Morphological changes involving synthesis, reorienting and lysis of the cell wall structure are an essential process in fungi (Enderlin and Selitrennikoff, 1994; Seiler and Plamann, 2003). The cell wall forms a multilayered complex of polysaccharides, glycoproteins and proteins with cross-linkages. The polysaccharides consist of β-glucans (mainly β-1,3-glucan and β-1,6-glucan), chitin, chitosan, mannans and α-glucans (Aimanianda et al., 2009; Fontaine et al., 2000; Kollár et al., 1997). Fungi such as ascomycetes and basidiomycetes produce enzymes associated with these polysaccharides. Some of these fungal glycoside hydrolases (GH) act on cell wall components and are responsible for morphological changes (Adams, 2004; Fukuda et al., 2008; Mahadevan and Mahadkar, 1970; Wessels and Niederpruem, 1967).

Most basidiomycetes form a fruiting body (mushroom) for sporulation. The cell walls of the fruiting body are constructed mainly from chitin, β-1,3-glucan and β-1,6-glucan. Several cell wall polysaccharides extracted from basidiomycetes, such as *Schizophyllum commune* (Tabata et al., 1990), *Agaricus blazei* (Ohno et al., 2001), *Coprinopsis cinerea* (Bottom and Siehr, 1979), *Grifola frondosa* (Ishibashi et al., 2001) and *L. edodes* (Chihara et al., 1969), show bioactive (antitumor) activity, and some of these antitumor compounds are β-1,3-glucans with β-1,6-linked branches, i. e., lentinan isolated from *L. edodes* (Chihara et al., 1969) and schizophyllan from *S. commune* (Ooi and Liu, 2000). Studies of these mushroom polysaccharides have indicated that their content decreases during storage after harvest, suggesting that polysaccharides of the cell wall are self-degraded by enzymes associated with cell wall autolysis during fruiting body senescence (Minato et al., 2004). Two types of exo-β-1,3-glucanases, EXG1 (Sakamoto et al., 2005a) and EXG2 (Sakamoto et al., 2005b), and two endo-β-1,3-glucanases, TLG1 (Sakamoto et al., 2006) and GLU1 (Sakamoto et al., 2011) have been reported in *L. edodes* fruiting bodies. EXG1 is classified in GH family 5 and EXG2 is classified in GH family 55. TLG1 is not classified in the GH family, so far, and TLG1 is similar to thaumatin-like protein (Sakamoto et al. 2006). GLU1 is classified in a new GH family, GH128 (Sakamoto et al. 2011).

β-1,6-glucan is thought to be a unique and essential component of fungal cell walls. Whereas β-1,3-glucan forms a microfibrillar structure, β-1,6-glucan forms a branched amorphous structure (Kollár et al., 1997). Lichen *Umbilicaria* species produce a linear glucan composed of only β-1,6-linkages (pustulan; Nishikawa et al., 1970). Many fungi are considered to secrete β-1,6-glucanases, some of which have been purified and characterized (Bryant et al., 2007; Oyama et al., 2002; Moy et al., 2002; Pitson et al., 1996). All of the isolated enzymes are β-1,6-glucan endohydrolases (EC 3.2.1.75) classified into GH family 5 or 30 of the GH-A clan in the CAZy database. Konno and Sakamoto (2011) first reported a β-1,6-glucanase in basidiomycetes, which is classified in the GH30 family, from *L. edodes* fruiting body.

3.1.1 exo-β-1,3-glucanase

Several exo-β-1,3-glucanases genes of fungi have been isolated and characterized. In *Saccharomyces cerevisiae,* three exo-β-1,3-glucanase encoding genes (*EXG1, EXG2, SSG1*) have been cloned and characterized (Larriba et al., 1995). The *EXG1* gene encodes two main extracellular exo-β-1,3-glucanases (Kuranda and Robbins, 1987; Vázquez de Aldana et al., 1991), *EXG2* encodes an exo-β-1,3-glucanase attached to the plasma membrane (Correa et al.,

1992; Larriba et al., 1995) and *SSG1* (also known as *SPR1*) encodes a sporulation-specific exo-β-1,3-glucanase (Muthukumar et al., 1993). Several other exo-β-1,3-glucanase genes have been isolated from yeasts, including *Candida albicans* (Chambers et al., 1993), *Kluyveromyces lacis*, *Hansenula polymorpha* and *Schwanniomyces occidentalis* (Esteban et al., 1999). In filamentous fungi, an exoglucanase encoding gene, *EXG2*, was isolated from the plant pathogenic fungus *Cochliobolus carbonum* (Kim et al., 2001). In basidiomycetous mushrooms, there is little information about exo-β-1,3-glucanases encoding genes such as a report of two open reading frames (ORF) for exo-β-1,3-glucanase sequences in *A. bisporus* (van den Rhee et al., 1996). Sakamoto et al., (2005b) purified and characterized an exo-β-1,3-glucanase from *L. edodes*, designated EXG1. EXG1 is a β-1,3-glucanase classified in the GH5 family. EXG1 has high similarity to exo-β-1,3-glucanase sequences in *A. bisporus* (van den Rhee et al., 1996). EXG1 in *L. edodes* is specifically expressed in fruiting bodies but not in vegetative mycelia, and decreases after harvesting. EXG1 can degrade a β 1,3-glucan, laminarin, but not lentinan (Sakamoto et al., 2005a; Table 1). These observations suggest that EXG1 is not involved in lentinan degradation or fruiting body senescence. On the other hand, EXG1 is expressed abundantly in growing stipes. Cell wall degrading and rearranging enzymes are important for stipe elongation; therefore, EXG1 could have a function in stipe elongation in *L. edodes* fruiting bodies.

	EXG1[1)]	EXG2[2)]	TLG1[3)]	GLU1[4)]	PUS30[5)]
cleavage type	exo	exo	endo	endo	endo
cleavage linkage	β-1,3	β-1,3	β-1,3	β-1,3	β-1,6
GH family	GH5	GH55	-	GH128	GH30
Lentinan (β-1,3-1,6)	× [a)]	○ [b)]	○	○	×
laminarin (β-1,3-1,6)	○	○	○	○	○
pusturan (β-1,6)	×	×	×	×	○
expression after harvesting	decrease	increase	increase	increase	increase

1) Sakamoto et al. 2005a; 2) Sakamoto et al. 2005b; 3) Sakamoto et al. 2006; 4) Sakamoto et al. 2011; 5) Konno and Sakamoto 2011
a) not degraded b) degraded

Table 1. Summary of glucanases (EXG1, EXG2, TLG1, GLU1, and PUS30) purified from *L. edodes*.

Several ascomycetous fungi, including *Aspergillus saitoi* (Oda et al., 2002), and the mycoparasitic fungi *Trichoderma harzianum* (Cohen-Kupiec et al., 1999) and *Ampelomyces quisqualis* (Rotem et al., 1999) express exo-β-13-glucanases other than the GH5 type of exo-β-1,3-glucanases. These exo-β-13-glucanases are classified in the GH55 family. An second exo-β-1,3-glucanase EXG2 was purified and characterized from *L. edodes* fruiting body after harvest (Sakamoto et al. 2005b). EXG2 is expressed in growing stipe but only weakly expressed in fresh gills. Expression of EXG2 increases immediately after harvest, and then becomes abundant in fruiting bodies 3 days after harvest. EXG2 has a high ability for degrading lentinan (Sakamoto et al. 2005b; Table 1), producing mainly glucose and gentiobiose (Fig. 2A). This suggests that EXG2 can hydrolyze β-1,3-glucan linkages in spite of the existence of a β-1,6-glucan linkage in lentinan (Fig. 2A). This hydrolyzing activity is demonstrated in a GH55 enzyme in *Phanerochaete chrysoporium* (Ishida et al., 2010). These observations suggest that EXG2 in *L. edodes* is mainly related to lentinan degradation after harvest. EXG2, like EXG1, is also abundantly expressed in growing stipes. This suggests that EXG2 has a dual function in stipe elongation and fruiting body senescence.

3.1.2 endo-β-1,3-glucanase

A. bisporus produce an endo-β-1,3-glucanase (Galán et al., 1999) and an endo-β-1,3-glucanase has also been reported from *L. edodes* (Grienier et al., 2000) that exhibits similarities to the anti-fungal thaumatin-like (TL) proteins that are highly conserved in plants. Plants accumulate a large number of pathogenesis-related (PR) proteins, which are divided into five families (PR1-PR5), and TL proteins share sequence homology with the thaumatin isoforms from arils of *Thaumatococcus danielli* (Dudler et al., 1994) are members of the PR5 family (van Loon and van Strien, 1999). Some TL proteins exhibit both β-1,3-glucan binding (Tundel et al., 1998), and endo-β-1,3-glucanase activities (Grenier et al., 1999). Proteins with endo-β-1,3-glucanase activities in mushrooms that have glucan binding activity were purified (Grienier et al., 2000), and their N-terminal amino acid sequences were similar to TL proteins. Sakamoto et al. (2006) purified an endo-β-1,3-glucanase, TLG1, and isolated the encoding gene *tlg1*, which has high similarity to TL proteins. TL protein encoding genes have been found in organisms outside of the plant kingdom, such as in the nematode *Caenorhabditis elegans* (Kitajima and Sato, 1999) and in the locust *Schistocerca gregaria* (Brandazza et al., 2004). It is reported that genes with similarity to TL proteins are highly conserved in fungi (Sakamoto et al., 2006). Fungal TL proteins also exhibit both β-1,3-glucan binding and β-1,3-endoglucanase activities (Grenier et al., 2000). Expression of TLG1 in *L. edodes* is significantly weak in vegetative mycelia, growing stipes and fresh fruiting bodies (Sakamoto et al. 2006). TLG1 is only expressed abundantly in fruiting bodies after harvest. TLG1 has lentinan and cell wall degrading activity (Sakamoto et al., 2006; Table 1), suggesting that TLG1 is specifically involved in lentinan degradation and fruiting body senescence.

GLU1 was purified as an endo-β-1,3-glucanase from *L. edodes* fruiting bodies after harvest, and was separated from TLG1. GLU1 showed greater hydrolyzing activity against laminarin than against lentinan by endo lytic manner (Fig. 2A; Sakamoto et al., 2011). The gene encoding GLU1 was isolated (*glu1*), and recombinant GLU1 in *Pichia pastoris* had endo- β-1,3-glucanase activity (Sakamoto et al., 2011). However, GLU1 did not have significant similarity to known β-1,3-glucanases or to any glycoside hydrolases. Endo- β-1,3-glucanases are divided into two classes, EC 3.2.1.6 [endo-1,3(4)- β-glucanase] and EC 3.2.1.39 (endo-1,3- β-glucanase), based mainly on substrate specificity. Because GLU1 did not degrade β-1,3-linkages within β-1,3-1,4-glucans such as barley glucan, the enzyme was categorized into EC 3.2.1.39. A phylogenetic tree was drawn based on sequences; GLU1 obviously does not belong to existing GH families containing EC 3.2.1.39 glucanases on the CAZy server (Sakamoto et al., 2011). Moreover, amino acid sequence analysis of GLU1 revealed no significant homology with any previously described functional proteins, the enzyme and other similar proteins were classified in a new GH family, GH128. Expression of *glu1* is weak in vegetative mycelia and growing fruiting bodies, but increases after harvest (Fig. 3). Its expression pattern is very similar to TLG1

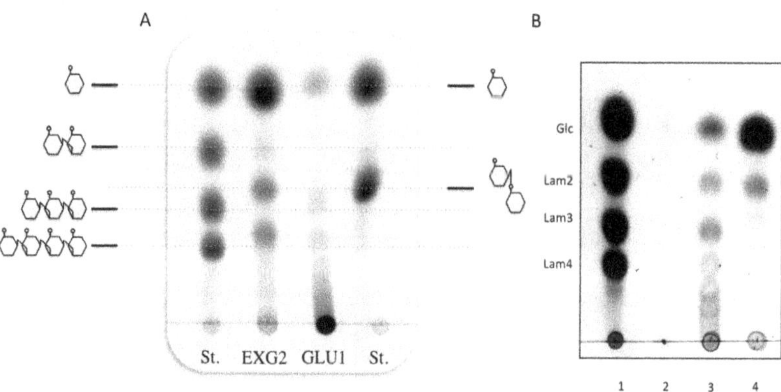

Fig. 2. A Degradation of β-1,3-1,6-glucan (lentinan) by EXG2 and GLU1. Left standards (St.) indicates β-1,3-glucan oligosaccharides, and right St. indicates β-1,6-glucan oligosaccharides. B Combination of GLU1 and EXG2 from *L. edodes* on the β-1,3-glucan hydrolysis. Pachyman (1%, w/v) (lane 2, untreated substarate) was firstly treated with GLU1 (lane3), and the GLU1-treated pachyman was further incubated with of EXG2 (lane 4). Standards (lane 1) are glucose (Glc), laminaribiose (Lam2), laminaritriose (Lam3) and Laminaritetraose (Lam4).

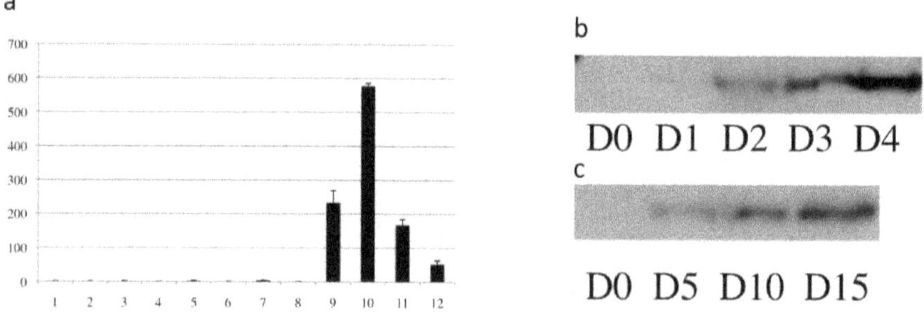

Fig. 3. Expression pattern of the *glu1* a: transcription levels of the *glu1*. 1: mycelium from liquid culture. 2: young fruiting bodies under 1 cm. 3: young fruiting bodies, 1-2 cm. 4: stipe of young fruiting bodies, 2-3 cm. 5: pileus of young fruiting bodies, 2-3 cm. 6: stipe of young fruiting bodies, 3-5 cm. 7: pileus of young fruiting bodies, 3-5 cm. 8: gill of mature fruiting body; 9: gill of fruiting body at 1 day after harvest; 10: gill of fruiting body at 2 days after harvest; 11: gill of fruiting body at 3 days after harvest; 12: gill of fruiting body at 4 days after harvest. b: Western blot analysis of GLU1 after harvest. D0, D1, D2, D3 and D4 indicate day 0 (fresh fruiting body), day 1 day 2, day 3, day 4 after harvesting, respectively. c: Western blot analysis of GLU1 after spore dispersal. D0, D5, D10, D15 indicates day 0, day 5 day 10, day 15 after cap veil open.

(Sakamoto et al. 2006). GLU1 can degrade lentinan, suggesting that GLU1 is involved in lentinan degradation after harvest. Enzymatic properties and expression pattern of GLU1 and

TLG1 are very similar; therefore, it is considered that these two enzymes have a redundant function in senescence of the *L. edodes* fruiting body.

It has been reported that some of the enzymes in the GH16 family have endo- β-1,3-glucanase activity. One GH16 family gene, *mlg1*, was found in fruiting bodies after harvest (Sakamoto et al., 2009). Furthermore, several other GH16 family genes, *mlg2* (DJ432070), *ghf16.1* (DJ432068), and *ghf16.2* (DJ432069) were isolated from *L. edodes* fruiting bodies. However, the enzymatic activities of the proteins encoded by these genes are not known; therefore, there is no clear evidence for a relationship between GH16 family enzymes and autolysis of the *L. edodes* fruiting body so far.

3.1.3 endo-β-1,6-glucanase

Some mycoparasitic fungi such as *Trichoderma* species produce an extracellular β-1,6-glucanase, member of GH30, for attack and degradation of host cell walls during their mycoparasitic action (De la Cruz and Llobell, 1999; Djonović et al., 2006; Montero et al., 2005). However, little information is known about the physiological function and role of the fungal β-1,6-glucanases. Sakamoto et al. (2009) found one GH30 protein, *ghf30* is up-regulated after harvest. Furthermore, LePus30A, an endo-type β-1,6-glucan hydrogenase, and classified as a member of GH family 30, was purified from *L. edodes* (Konno and Sakamoto 2011). LePus30A was the first purified basidiomycetous protein characterized as a GH 30 member (Konno and Sakamoto 2011). LePus30A has high levels of similarity to proteins from basidiomycetous species such as *L. bicolor*, *S. commune* and *C. cinerea*, suggesting that β-1,6-glucanases widely conserved in basidiomycetous fungi. The transcript level of *lepus30a* in fruiting bodies undergoing postharvest preservation for 2-4 days was also significantly higher than at other stages of the life cycle. This result supported an important role for LePus30A in the degradation of the cell wall's complex structure during fruiting body senescence after harvest. LePus30 has no activity toward lentinan, but the enzyme showed activity for the cell wall glucans from *L. edodes* fruiting bodies (Konno and Sakamoto 2011). In addition, LePus30A degrades cell wall glucans producing glucose and β-1,6-linked oligoglucosides. Therefore, LePus30A is mainly contribute to a degradation of the β-1,6-glucan rich content, during fruiting body senescence. The expression of *lepus30a* was also observed in mycelium and young fruiting body. This result implicates that the β-1,6-glucanase also contributes in hyphal growth and branching, or development (Djonović et al. 2006; Moy et al. 2002).

3.1.4 lentinan degradation

Cell wall degrading enzymes, namely exo-β-1,3-glucanase EXG2 and endo-β-1,3-glucanases TLG1 and GLU1 are involved in lentinan degradation after harvesting (Table 1). To study the cooperation of *L. edodes* endo- and exoglucanases during β-1,3-glucan hydrolysis, pachyman (a linear β-1,3-glucan from the basidiomycete *Poria cocos*) hydrolysates obtained after treatment with GLU1 were further treated with EXG2, and the products were analyzed (Fig. 2B). As a result, formed various length of oligosaccharides through the endo-type manner of GLU1 were further degraded to glucose and disaccharide (laminaribiose) by the exo-type manner of EXG2 (Fig. 2B). Thus, the cell wall after harvesting of *L. edodes* fruiting bodies appear to be degraded by combination of endo-type glucanases (GLU1 and TLG1), and exo- β-1,3-glucanase (EXG2). GLU1 and TLG1 were suggested to have similar biological functions in the process of lentinan degradation after harvest. LePus30A showed its high substrate specificity for a linear β-1,6-glucan polysaccharide (pustulan) and very low activity toward a branched β-1,3/1,6-glucan (laminarin). Since mainly glucose was liberated when laminarin was treated with LePus30A, the enzyme presumably cut the β-1,6-linked side chains in laminarin (DP range 20-30, β-1,3/1,6=7/1). TLG1 and GLU1 shorten main chain length of lentinan, therefore, side chain of partially degraded lentinan would be cut by PUS30A. After all, lentinan is degraded immediately after harvest by cooperation of increased EXG2, TLG1, GLU1, PUS30A after harvest to oligo-saccharides and glucose (Fig. 4).

3.2 Chitinase

As chitin is one of the fungal cell wall components, chitinase and chitin synthase are very important for morphogenesis of fungi. There are several reports on chitinase activities in basidiomycetous mushrooms (for example, Kamada et al. 1981), but few on the relationship between chitinase and fruiting body senescence. Higher chitinase activity is observed in fruiting bodies following harvest compared to the activity just after harvest (Fig. 5). Genes encoding chitin degrading enzymes were identified in fruiting bodies after harvest (*chi1*,

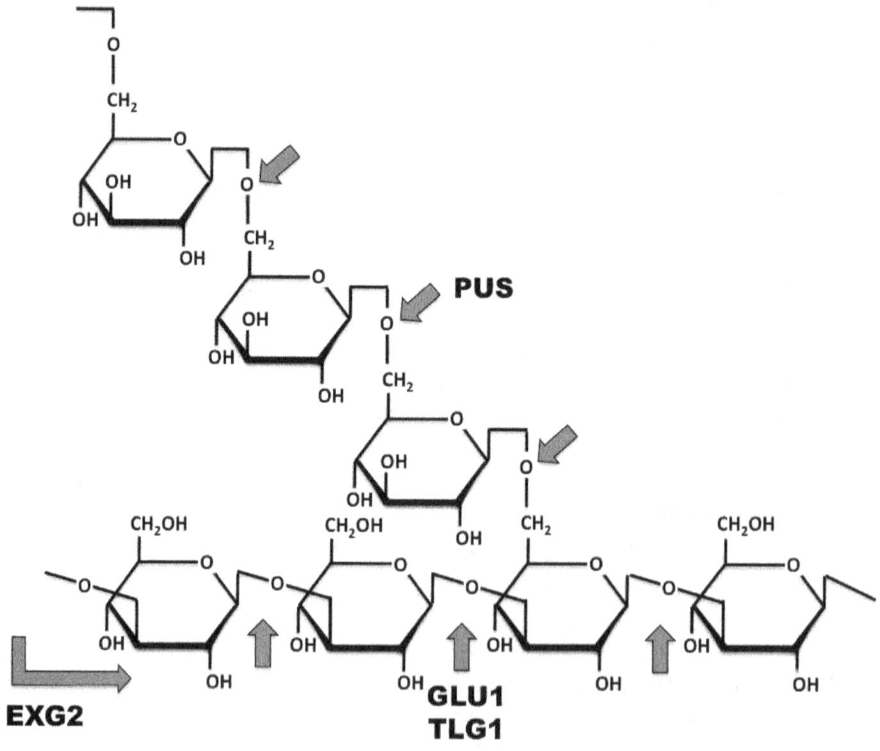

Fig. 4. Scheme of degradation of β-1,3-1,6-glucan after harvest of the *L. edodes* fruiting body

chi2; Sakamoto et al., 2009). The putative amino acid sequences of *chi1* and *chi2* contain a motif found in the GH18 family (Fig. 6). These genes do not have significant similarity to known chitinase encoding genes, but have significant similarities to hypothetical genes in basidiomycetous genomes such as *S. commune, Serpula lacrymans* and *Postia placenta.*

A putative chitinase encoding gene has also been found in an EST sequence from *L. edodes* (Suizu et al. 2008); the cloned full-length gene, *chi3*, has a GH18 domain and chitin binding domains (Fig. 6). Expression of *chi3* also increased after harvest (Fig. 7). These observations suggest that increased expression of chitinases has an important role in fruiting body

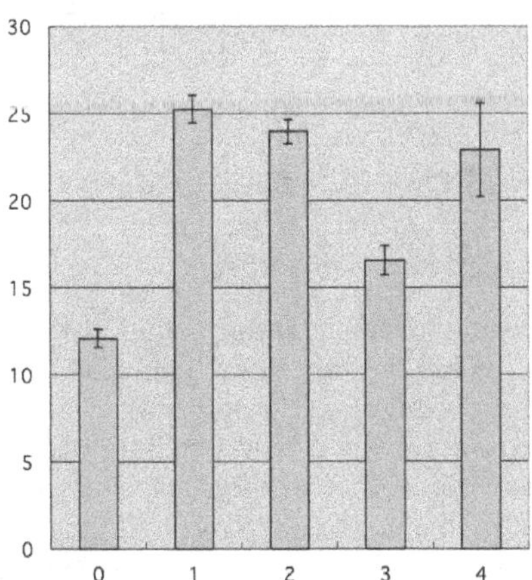

Fig. 5. chtinase activity after harvest measured by using 4-MUF-chitin (Hood 1991). The X axis indicates 0: fresh fruiting body; 1: fruiting body at day 1 after harvest ; 2 fruiting body at day 2 after harvest; 3 fruiting body at day 3 after harvest; 4: fruiting body at day 4 after harvest. The Y axis indicates unit/μg protein.

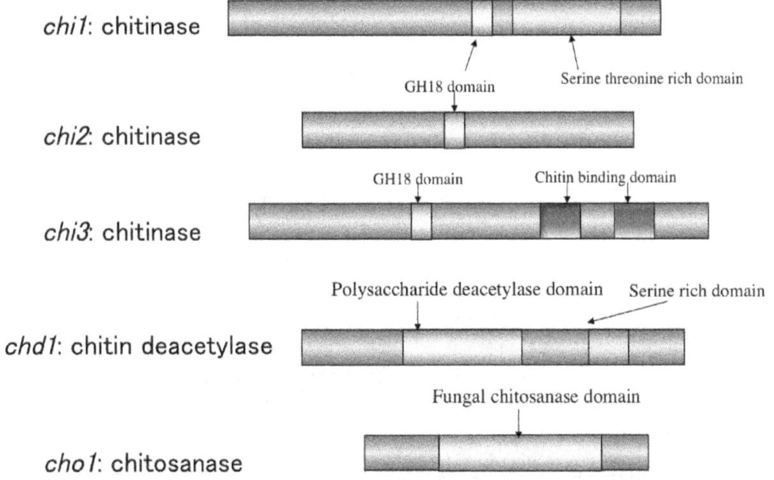

Fig. 6. Chitin related enzymes upregulated after harvest of the *L. edodes* fruiting body.

senescence after harvest. Other genes encoding putative enzymes related to chitin modification were cloned from fruiting bodies after harvest (Sakamoto et al., 2009), including chitin deacetylase (*chd1*) and chitosanase (*cho1*). The putative amino acid sequence of *chd1* has a polysaccharide deacetylase domain in the middle and a serine-rich region at the C-terminus that are present in the chitin deacetylase from *C. neoformans* (Levitz et al., 2001; Fig 6). The DNA sequence of the chitosanase from *L. edodes, cho1,* has high similarity with that of *A. oryzae,* but there are no significantly similar sequences in the basidiomycetous genome databases available so far. Chitosan is also a cell wall component of *L. edodes* (Pochanavanich et al., 2002). It was reported that chitin deacetylase (*chd1*) and chitosanase (*cho1*) were upregulated after harvesting of the *L. edodes* fruiting body (Sakamoto et al., 2009). The observation that the *chi1, chi2* and *cho1* genes do not have significant similarity to other basidiomycetous genes suggests that *L. edodes* has a unique chitin and chitosan metabolism system. These data indicate that the cell wall of the *L. edodes* fruiting body is possibly degraded as a result of increased glucanase and chitinase activity following harvest.

4. Regulation of senescence related genes

As shown above, many genes involved in fruiting body senescence increased after harvest of the *L. edodes* fruiting body, including those for phenol oxidases, involved in gill browning, and cell wall degrading enzymes involved in autolysis after harvest. Expression of many other genes increased after harvest of the *L. edodes* fruiting body (Sakamoto et al., 2009). The functions of these genes in senescence of the fruiting body are still unclear, but certainly gene expression is drastically changed after harvest. Transcription factors or chromatin remodelling could cause these drastic changes in gene expression. A gene encoding a putative transcription factor, *exp1,* was isolated from *L. edodes* fruiting bodies after harvest. The putative amino acid sequence of *exp1* displays high similarity with the sequence for *exp1* from *C. cinerea* (Accession No. AB363984; Muraguchi et al., 2008), and contains two HMG boxes in the C-terminus. Proteins that have HMG boxes have functions in transcription factor or chromatin remodeling. In this section, genes other than phenol oxidases and cell wall degradation enzymes, but expressed after harvest will be discussed.

4.1 Unknown genes increased after harvesting

Expression of numerous genes increases after harvest of *L. edodes* fruiting bodies, but their functions in senescence are still unknown. For example, genes encoding riboflavin forming enzyme (*baw28*) and malate dehydrogenase (*mdh*) were isolated from *L. edodes* fruiting bodies. The putative amino acid sequence of *baw28* gene displays 53% identity to the *C. albicans* riboflavin forming enzyme and 33% identity to the *L. edodes* riboflavin forming enzyme (Accession No. AB116639) that is expressed specifically in the fruiting body (Hirano et al., 2004). The putative amino acid sequence of *baw28* has a barwin-like endoglucanase domain in the N-terminus, and a serine-rich region in the C-terminus. The *mdh* gene identified in this study does not contain any significant motifs, but displays 34% identity to *Aspergillus fumigatus* malate dehydrogenase. Expression of the genes increased after harvest (Fig. 7); therefore, the genes presumably have some biological function in senescence, but their function is unclear.

Many genes upregulated after harvest in the *L. edodes* fruiting body have been reported but do not have significant similarities to other genes (Sakamoto et al., 2009; Fig. 8). On the other hand, genome sequence information in basidiomycetous fungi is increasing, so extensive new information has been found from unknown genes. One of them, a gene that has similarity to glucoamylase (Fig. 8 *amy*). Expression of *amy1* is weak in vegetative mycelia and young fruiting bodies, but significantly increases after harvest. The *ghf79* has similarity to GH79 family which includes enzymes glucronidase. The gene *lup33* has high similarity to hypothetical proteins found in basidiomycetous genomes. The function of the hypothetical proteins is unclear, but expression of the gene specifically increases in the *L. edodes* fruiting body after harvest. On the other hand, there are several genes that have no similarity to any other genes, including hypothetical proteins in the basidiomycetous genomes available so far. These genes are *L. edodes* specific genes. Expression of these *L. edodes unknown protein* encoding genes, *lup23, lup410, lup66,* and *lup48,* increases significantly after harvest (Sakamoto et al., 2009, Fig. 8), but is weak in vegetative mycelia and young fruiting bodies. The function of these genes is still unknown, but these genes would have a specific role in senescence of the *L. edodes* fruiting body after harvest.

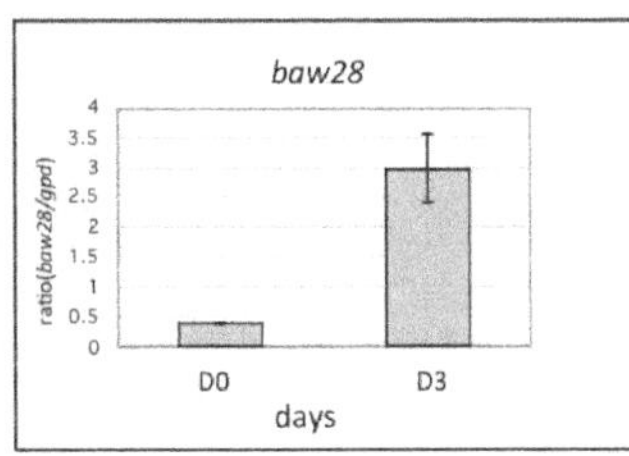

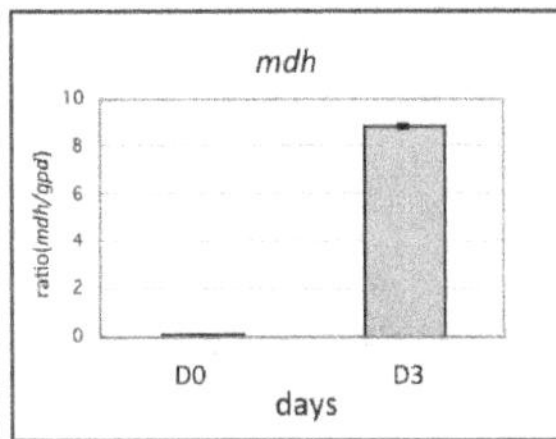

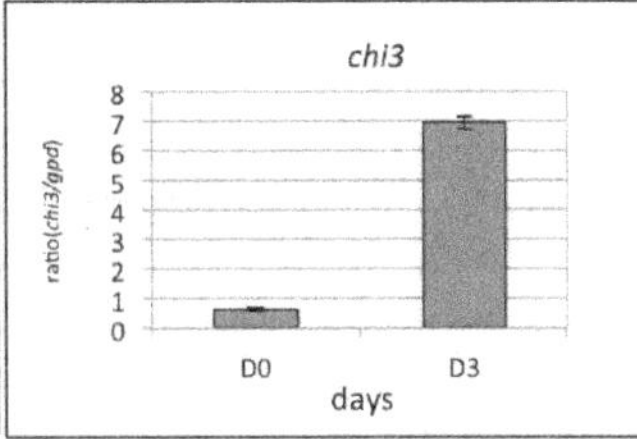

Fig. 7. Comparison of expression levels of *baw28, mdh* and *chi3* in fresh fruiting bodies and fruiting bodies (D0) at day 3 (D3) after harvest. The Y axis represents the ratio of mRNA levels of each gene to that of *gpd*.

4.2 putative senescence related gene transcription regulating factor *exp1*

Putative transcription factors, *exp1* which is up-regulated after harvest are isolated from *L. edodest*. As the mRNA level of *exp1* was higher three days after harvest than on the day of harvest, *exp1* is likely to be involved in fruiting body senescence after harvesting. The *L. edodes exp1* gene, a homolog of *exp1* isolated from *C. cinerea* (Muraguchi et al., 2008), contains two HMG boxes in the C-terminus. The HMG1/2 class proteins have been considered architectural components of chromatin that have a general role in the regulation of chromosomal functions (Thomas et al., 2001). Several proteins that have an HMG box are considered to be transcription factors, or have been shown to interact with transcription factors (Wissmüller, 2006).

C. cinerea opens the cap of its fruiting body by lysing lamellae in the cap during spore diffusion, and the *C. cinerea exp1* mutant cannot open its cap (Muraguchi et al., 2008). This suggests that *exp1* in *C. cinerea* controls cap autolysis during spore diffusion by regulating the genes that encode cell wall lysing enzymes. For example, expression of *tlg1,* which is a homolog of one of the cell wall degrading enzymes for senescence in *L. edodes,* is suppressed

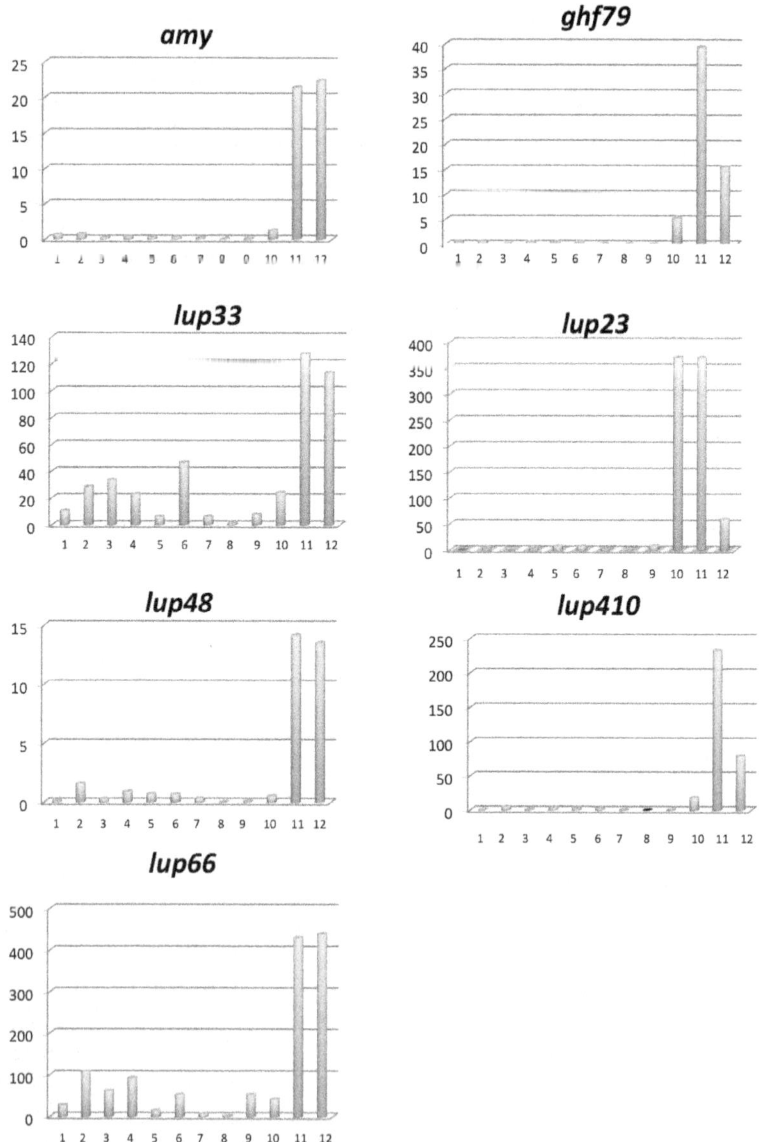

Fig. 8. Expression pattern of unknown genes found in the *L. edodes* fruiting body. The Y axis represents the ratio of mRNA levels of each gene to that of *gpd*. The X axis indicates 1: mycelium from liquid culture. 2: young fruiting bodies under 1 cm. 3: young fruiting bodies, 1-2 cm. 4: stipe of young fruiting bodies, 2-3 cm. 5: pileus of young fruiting bodies, 2-3 cm. 6: stipe of young fruiting bodies, 3-5 cm. 7: pileus of young fruiting bodies, 3-5 cm. 8: gill of mature fruiting body; 9: gill of fruiting body at 1 day after harvest; 10: gill of fruiting body at 2 days after harvest; 11: gill of fruiting body at 3 days after harvest; 12: gill of fruiting body at 4 days after harvest

in the *C. cinerea exp1* mutant (Fig. 9). Cell wall lysis after spore diffusion is observed in *L. edodes*, and expression of the cell wall degrading enzymes EXG2, TLG1 and GLU1 increases after spore diffusion (Sakamoto et al., 2005b, 2006, Fig. 3). Expression of these cell wall degrading enzymes also increased after harvest, suggesting that the systems for cell wall lysis in the fruiting body after harvesting and after spore diffusion are similar in *L. edodes*. These results suggest that *exp1* in *L. edodes* might control senescence of the fruiting body by regulating genes that are expressed after harvesting, such as genes that encode cell wall degrading enzymes.

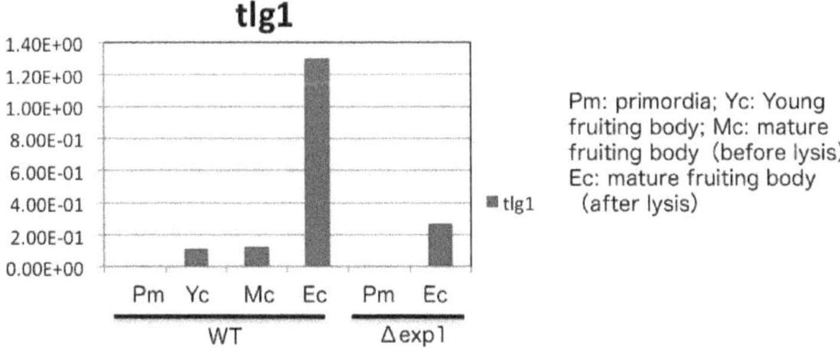

Fig. 9. expression pattern of the *tlg1* in Δexp1. The Y axis represents the ratio of mRNA levels of the *tlg1* to that of *gpd*.

5. Genes downregulated after harvesting

It has been reported that a large number of genes downregulated after harvest are likely involved in normal fruiting body formation and are needed to maintain freshness of the fruiting body (Sakamoto et al. 2009). Translation, transcription and protein metabolism related genes, in addition to those involved in spore formation, cytoskeleton, and cell cycle are downregulated after harvest, and many more transcription factors are also downregulated after harvest (Sakamoto et al. 2009). Interestingly, genes identified among those upregulated after harvest are completely different from those in the downregulated genes. This suggests that gene transcription is drastically altered after harvesting of the fruiting body.

5.1 Transcription, translation, and protein metabolism related genes

Many transcription, translation and protein metabolism related genes are down-regulated after harvesting. Several RNA related genes downregulated after harvest, such as the Pumilio family RNA binding protein, RNA binding protein 5-like protein, ATP-dependent helicase and pre-mRNA splicing factor (Sakamoto et al., 2009). These RNA related proteins are involved in transcription and translation (de Moore et al., 2005; Rogers et al., 2002); therefore, expression of many genes might be changed after harvesting as a result of suppression of these transcription and translation related genes. Heat shock proteins (i.e., Hsp70 and Hsp90), chaperonin and calnexin, which are involved in the proper translation and folding of proteins to control protein quality (Saibil, 2008; Spiess et al., 2004; Williams, 2006), were also down-regulated. This suggests that the protein quality control system in the

L. edodes fruiting body is less effective following harvest. Hsp70 and Hsp90 were especially highly expressed in young fruiting bodies (Fig. 10); therefore, these proteins have an important function in normal fruiting body development. Several proteases and peptidases, such as aspartic protease (*pro1, pep1*), metallopeptidase MepB and gamma-glutamyltranspeptidase were downregulated after harvest (Sakamoto et al., 2009), and proteasome-related protein encoding genes such as proteasome 26S, proteasome 26S subunit alpha type 6 and proteasome 26S ATPase subunit 3 were also downregulated after harvest (Sakamoto et al., 2009). These

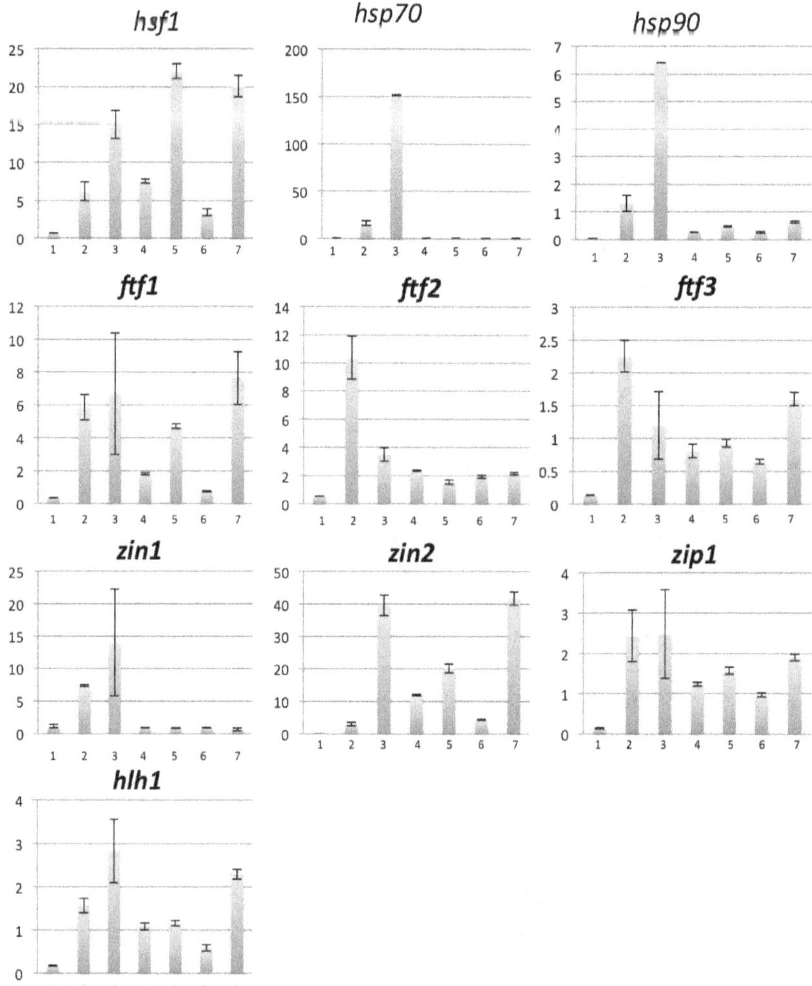

Fig. 10. Putative transcription factors found in the *L. edodes* fruiting body. The Y axis represents the ratio of mRNA levels of each gene to that of *gpd*. The X axis indicates 1: mycelium from liquid culture. 2: young fruiting bodies under 1 cm. 3: young fruiting bodies, 1-2 cm. 4: stipe of young fruiting bodies, 2-3 cm. 5: pileus of young fruiting bodies, 2-3 cm. 6: stipe of young fruiting bodies, 3-5 cm. 7: pileus of young fruiting bodies, 3-5 cm.

proteins comprise a proteasome complex that is involved in protein degradation (Wolf and Hilt, 2004). Decreased protease and proteasome activity would extend the life of proteins expressed after harvest. These data suggest that transcription, translation and posttranslational protein metabolism are drastically altered in fruiting bodies after harvest.

5.3 Mitosis and meiosis related genes

Many mitosis and meiosis related genes are down-regulated after harvesting. Several meiosis related proteins, such as a meiotic recombination related protein encoding gene, *dmc1*, show altered expression in harvested fruiting bodies (Sakamoto et al., 2009). DMC1 is a recombinase that is required for DNA pairing during recombination. Two DMC1-encoding genes were previously identified in *C. cinerea*, namely Rad51 (Stassen et al., 1997) and LIM15 (Namekawa et al., 2005). Whereas Rad51 is involved in both mitosis and meiosis, LIM15 is specifically involved in meiosis, being expressed during meiosis and disappearing immediately after meiosis (Nara et al., 1999). The putative amino acid sequence encoded by *dmc1* has significant similarity to LIM15/DMC1 in *C. cinerea*. The *dmc1* gene was found to be transcribed abundantly in the gills of fresh fruiting bodies and was downregulated after harvesting (Sakamoto et al., 2009). These data suggest that *dmc1* is specifically involved in meiosis, and is downregulated after spore formation. Expression of the cell division related genes septin and *cdc48* reportedly decreased after harvest (Sakamoto et al., 2009). Both of these genes have important functions during cell division (Lindsey and Momany, 2006; Cheeseman and Desai, 2004). Septins are GTPases that form filaments in fungi and animals and are involved in membrane trafficking, coordinating nuclear divisions, and organizing the cytoskeleton (Lindsey and Momany, 2006). cdc48 has a variety of cellular functions; for instance, it is implicated in the membrane fusions that occur after mitosis to reassemble the endoplasmic reticulum and the Golgi apparatus (Cheeseman and Desai, 2004). Genes encoding the cytoskeleton related proteins actin and beta-tubulin were also downregulated after harvest (Sakamoto et al., 2009). These genes are essential for progression of cytokinesis (Nanninga, 2001). This suggests that the frequency of cell division decreases after harvest.

5.4 Putative transcription factors identified by reverse subtraction

Several putative transcription factors are downregulated after harvest of the *L. edodes* fruiting body. One putative transcription factor gene, *hlh1*, which contains a basic helix-loop-helix motif involved in DNA binding (Fig. 11), was also cloned (Sakamoto et al., 2009). Basic helix-loop-helix proteins are a group of eukaryotic transcription factors that exert a determinative influence on a variety of developmental pathways (Littlewood and Evan, 1995). The putative amino acid sequence of *hlh1* contains a nuclear localization signal and displays high similarity to a hypothetical protein from *C. cinerea*. The *hlh* is specifically expressed in young fruiting bodies (Fig. 10). Expression of *ftf1*, *ftf2*, *ftf3*, which have a fungal specific transcription factor domain (Fig. 11), was observed in *L. edodes* fruiting bodies (Sakamoto et al., 2009). The putative amino acid sequences of these genes have a similar structure to the *priB* gene product, which is involved in fruiting body development in *L. edodes* (Miyazaki et al., 2004). The *priB* gene is transcribed specifically in the *L. edodes* fruiting body, and regulates fruiting body-specific genes (Kaneko et al., 2001; Miyazaki et al., 2004).

The *ftf1 ftf2,* and *ftf3* are highly transcribed in fresh fruiting bodies and their mRNA levels decreased after harvesting (Fig. 10; Sakamoto et al. 2009), suggesting that the genes regulates genes involved in fruiting body development, but does not regulate genes involved in fruiting body senescence.

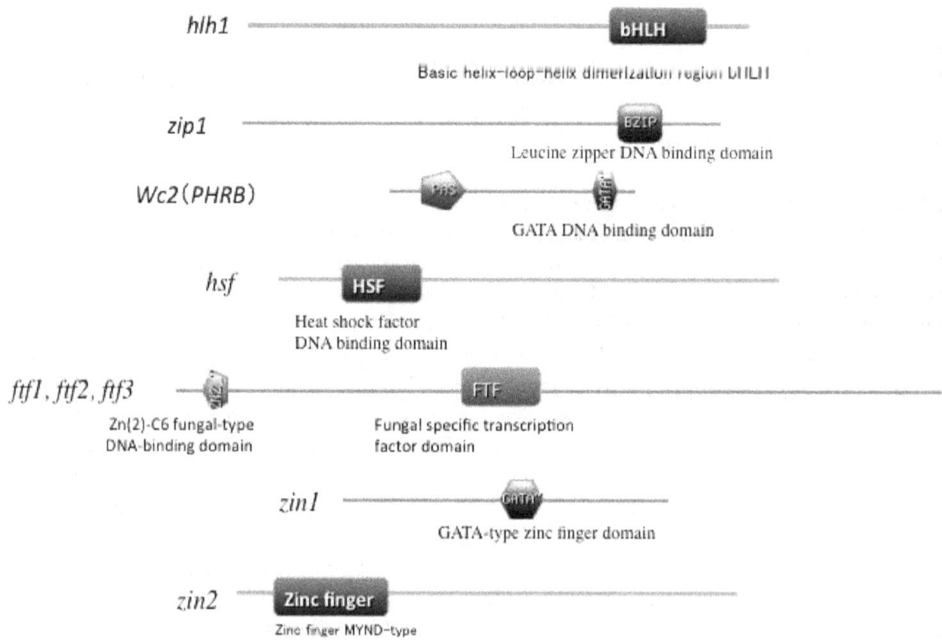

Fig. 11. Putative transcription factors found in the *L. edodes* fruiting body

The putative protein product of *hsf1* has no significant similarity to other known proteins, but has significant similarities to hypothetical proteins in several basidiomycetous fungi, such as *C. cinerea* and *Serpula lacryman*. The putative amino acid sequence of *hsf1* has a helix-turn-helix DNA binding domain for binding to a heat shock element (Fig. 11) that is conserved in the upstream region of heat shock proteins (Pirkkala et al., 2001). Its expression pattern was different from heat shock proteins (Fig. 10); therefore, *hsf* might not directly regulate the expression of heat shock proteins. The putative transcription factor, *zin1* and *zin2*, were also highly transcribed in fresh fruiting bodies, with a decline in expression after harvesting (Fig. 10; Sakamoto et al. 2009). The *zin1* and *zin2* gene products contain a zinc finger motif (Fig. 11). The putative amino acid sequence of *zin2* has a MYND-type zinc finger motif as well as significant similarity with the zf-MYND domain containing protein in *L. bicolor* that is upregulated by infection with ectomycorrhiza (Martin et al., 2008), but there is no significant similar sequence to the *zin2* gene sequences in other basidiomycetous genomes available so far. In contrast, there are other basidiomycetous proteins that have a MYND domain; for example, Fuz1 from *U. maydis* contains an MTND domain and is involved in cell morphogenesis (Chew et al., 2008). These results suggest that *zin2* is involved in fruiting body development. *zip* is also specifically expressed in fruiting bodies but not in vegetative mycelia (Fig. 10). The putative amino acid sequence of *zip1* has a leucine zipper DNA sequence, which is found in several transcription factors, and also has an Aft osmotic stress domain (Fig. 11). The *zip* does not have any significant similarity to known transcription factors but has significant similarities to hypothetical proteins in several basidiomycetous fungi, such as *C. cinerea* and *S. lacrymans*.

A homolog of the blue light receptor white collar 2 (WC2) in *N. crassa* was also found in *L. edodes* fruiting bodies. In *N. crassa*, WC2 acts as a component of the blue light receptor by interacting with white collar protein 1 (WC1). WC2 in *L. edodes* was cloned and characterized; the protein has been designated PHRB; Sano et al., 2009). Genes similar to WC1 have also been reported in *L. edodes* (PHRA; Sano et al., 2008) and in *C. cinerea* (*dst1*; Terashima et al., 2005). The *C. cinerea dst1* mutant cannot form a mature cap under normal light/dark conditions, and the shape of the fruiting body of the *dst1* mutant is very similar to that formed by the wild-type mushroom when grown in complete darkness (Terashima et al., 2005). Since WC2 and PHRA in *L. edodes* interact with each other, they could act as a blue light receptor in *L. edodes* that regulates fruiting body development in the light (Sano et al., 2010). Gene disruptionof the WC2 homolog in *C. cinerea* results in a similar phenotype to that of *dst1* (Nakazawa et al. 2011).

These genes might not be directly involved in senescence of the *L. edodes* fruiting body after harvest, but these putative transcription factor genes presumably would be involved in normal fruiting body morphology. A decrease in expression of the putative transcription factor genes would influence morphology after harvest.

6. Future perspectives

As shown above, numerous genes are related to senescence of the *L. edodes* fruiting body after harvest. However, the mechanism of senescence of the *L. edodes* fruiting body after harvest is not fully understood. To understand the mechanism involved, an understanding of total gene expression changes after harvest is needed. Transcriptome analysis such as microarray and SAGE analysis is useful to understand total gene expression changes. In *Lentinula edodes*, there are several reports of transcriptome analysis, such as an EST study (Suizu et al., 2009) and SAGE analysis (Chem 2011). For SAGE analysis, genomic sequence data are needed to increase accuracy. Public genomic sequence data of *L. edodes* are not yet available so far, but genomic sequencing will be easier than before using a next-generation genome sequencer, such as the Illumina GA IIx. Furthermore, SAGE analysis is optimized for next-generation sequencing to obtain a larger amount of sequence tagged data (Matsumura et al. 2003). Therefore, gene expression profile changes after harvesting of the *L. edodes* fruiting body will be revealed in more detail by using a combination of genome sequence analysis and Super-SAGE (Matsumura et al. 2003) with a next-generation sequencer. In *C. cinerea,* autolysis after spore dispersion is similar to that of *L. edodes*, and public genomic sequence data of *C. cinerea* are available, so it is possible to compare gene expression profiles during autolysis in *C. cinerea* and *L. edodes*. Such research will provide a basic understanding of mushroom senescence.

To prove the function of genes, gene disruption or gene silencing studies will be needed. Gene disruption systems have been constructed in several mushrooms such as *C. cinerea* (Nakazawa et al. 2011). A homologous gene recombination system has also been constructed (Irie et al. 2003), but a gene disruption system by homologous recombination has not yet been constructed in *L. edodes*. However, gene silencing by RNAi has succeeded in *L. edodes* (Nakade et al., 2011). Therefore, genes upregulated after harvest will be knocked down by RNAi, which will reveal the functions of the genes, such as *exp1* and *exg2*, in senescence. This research will provide useful data for breeding strains with fruiting bodies that remain fresh for a longer period after harvest.

7. Conclusion

After harvesting of the *L. edodes* fruiting body, drastic gene expression changes occur. In particular, expression of phenol oxidases (Tyrs and Lccs) and cell wall enzymes (glucanases and chitinases) increase after harvest. This suggests that a gene regulation system for senescence exists in *L. edodes*. *exp1*, which is a putative transcription factor or chromatin remodeling related protein, is one of the candidates for regulation of the drastic gene expression changes involved in senescence. The process would include programmed cell death, but there are few studies on the relationship between programmed cell death and mushroom morphology. Therefore, studies on senescence in mushrooms will be important not only for applied agricultural science, but also for basic science.

8. Acknowledgment

The authors thank Dr. Kamada, Okayama University and Dr. Muraguchi, Akita prefectural University for giving us the *exp1* mutant and related information in *C. cinerea*. The authors also thank to previous members of Microorganism lab. in Iwate biotechnology Research Center.

9. References

Adams D.J. (2004) Fungal cell wall chitinases and glucanases. Microbiology 150:2029–2035

Aimanianda, V.; Clavaud, C.; Simenel, C.; Fontaine, T.; Delepierre, M. & Latgé, J.P. (2009) Cell wall β-(1,6)-glucan of *Saccharomyces cerevisiae*: structural characterization and in situ synthesis. J Biol Chem 284:13401–13012

Aramayo, R.; & Timberlake, W.E. (1990) Sequence and molecular structure of the *Aspergillus nidulans* yA (laccase I) gene. Nucleic Acids Res 18, 3415.

Baldrian, P. (2006) Fungal laccases-occurrence and properties. FEMS Microbiol Rev 30: 215-242

Bottom, C.B. & Siehr, D.J. (1979) Structure of an alkali-soluble polysaccharide from the hyphal wall of the basidiomycete *Coprinus macrorhizus* var. *microsporus*. Carbohydr Res 77:169–181

Brandazza, A.; Angeli, S.; Tegoni, M.; Cambillau, C. & Pelosi, P. (2004) Plant stress proteins of the thaumatin-like familiy discovered in animals FEBS Lett 572: 3-7

Bryant, M.K.; May, K.J.; Bryan, G.T. & Scott, B. (2007) Functional analysis of a □-1,6-glucanase gene from the grass endophytic fungus *Epichloë festucae*. Fungal Genet Biol 44:808-817

Burton, K.S. (1988) The effects of pre- and post-harvested development on mushroom tyrosinase. J Hortic Sci 63, 255–260.

Chambers, R.S.; Broughton, M.J.; Cannon, R.D.; Carne, A.; Emerson, G.W. & Sullivan, P.A. (1993) An exo-β-(1,3)-glucanase of *Candida albicans*: purification of the enzyme and molecular cloning of the gene. J Gene Microbiol 139: 325-334

Cohen-Kupiec, R.; Broglie, K.E.; Friesem, D.; Broglie, R. & Chet, I. (1999) Molecular characterization of a novel β-1,3-exo-glucanase related to mycoparasitism of *Trichoderma harzianum*. Gene 226: 147-154.

Correa, J.; Vazquez de Aldana, C.R.; Segundo, P.S. & del Rey F (1992) Genetic mapping of 1,3-□-glucanase-encoding genes in *Saccharomyces cerevisiae*. Curr Genet 22: 283-288

Cheeseman, I. M., and Desai, A. 2004. Cell division: AAAtacking the mitotic spindle. Curr. Biol. 14: R70-72.

Chew, E., Aweiss, Y., Lu, C. Y., & Banuett, F. (2008) Fuz1, a MYND domain protein, is required for cell morphogenesis in *Ustilago maydis*. Mycologia. 100: 31-46.

Chihara, G.; Maeda, Y.; Hamuro, J.; Sasaki T. & Fukuoka, F. (1969) Inhibition of mouse srcoma 180 by polysaccharides from *Lentinus edodes* (Berk.) Sing. Nature 222: 687-688

Chum, W.W.; Ng, K.T.; Shih, R.S.; Au, C.H. & Kwan, H. S. (2008) Gene expression studies of the dikaryotic mycelium and primordium of *Lentinula edodes* by serial analysis of gene expression. Mycol. Res. 112: 950-964.

Chum, W.W,; Kwan, H.S.; Au, C.H.; Kwok, I.S. & Fung, Y.W. (2011) Cataloging and profiling genes expressed in *Lentinula edodes* fruiting body by massive cDNA pyrosequencing and LongSAGE. Fungal Genet Biol. Apr;48(4):359-69

Eastwood, D.C.; Kingsnorth, C.S.; Jonesa, H.E. & Burton, K.S. (2001) Genes with increased transcript levels following harvest of the sporophore of *Agaricus bisporus* have multiple physiological roles, Mycol Res, 105, 1223-1230

Eastwood, D.C.; Challen, M.P.; Zhang, C.; Jenkins, H.; Henderson, J. & Burton, K.S. (2008) Hairpin-mediated down-regulation of the urea cycle enzyme argininosuccinate lyase in *Agaricus bisporus*. Mycol Res 112: 708-716

Enderlin, C.S.; & Selitrennikoff, C.P. (1994) Cloning and characterization of a *Neurospora crassa* gene required for (1,3)β-glucan synthase activity and cell wall formation. Proc Natl Acad Sci USA 91:9500−9504

Esteban, P.F.; Vazquez, De Aldana, C.R. & Del Rey, F. (1999) Cloning and characterization of 1,3-β-glucanase-encoding genes from non-conventional yeasts. Yeast 15: 91-109

Espín, J.C.; Jolivet, S. & Wichers, H.J. (1999). Kinetic study of the oxidation of c-L-glutaminyl-4-hydroxybenzene catalyzed by mushroom (*Agaricus bisporus)* tyrosinase. J Agric Food Chem 47, 3495–3502.

De la Cruz, J. & Llobell, A. (1999) Purification and properties of a basic endo-β -1,6-glucanase (BGN16.1) from the antagonistic fungus *Trichoderma harzianum*. Eur J Biochem 265:145−151 de Moor, C.H.; Meijer, H. & Lissenden, S. (2005) Mechanisms of translational control by the 3' UTR in development and differentiation. Semin. Cell. Dev. Biol. 16: 49-58.

Djonović, S.; Pozo, M.J. & Kenerley, C.M. (2006) Tvbgn3, a β-1,6-glucanase from the biocontrol fungus *Trichoderma virens,* is involved in mycoparasitism and control of *Pythium ultimum*. Appl Environ Microbiol 72:7661–7670

Dudler, R.; Mauch, F. & Reimmann, C. (1994) Thaumatin-like protein. In M Witty, JD Higginbotham, eds, Thaumatin. CRC Press, Boca Raton, pp193-199.

Fontaine, T.; Simenel, C.; Dubreucq, G.; Adam, O.; Delepierre, M.; Lemoine, J.; Vorgias, C.E.; Diaquin, M. & Latgé, J.P. (2000) Molecular organization of the alkali insoluble fraction of *Aspergillus fumigatus* cell wall. J Biol Chem 275:27594–27607

Fukuda, K.; Hiraga, M.; Asakuma, S.; Arai, I.; Sekikawa, M. & Urashima, T. (2008) Purification and characterization of a novel exo-β-1,3-1,6-glucanase from the fruiting body of the edible mushroom Enoki (*Flammulina velutipes*). Biosci Biotechnol Biochem 72:3107–3113

Grenier, J.; Potvin, C.; Trudel, J. & Asselin, A: Some thauma- tin-like proteins hydrolyse polymeric β-1,3-glucans, Plant J, 19, 473-480 (1999)

Grenier, J.; Potvin, C. & Asselin, A. (2000) Some fungi express β-1,3-glucanases similar to thaumatin-like proteins. Mycologia 92: 841-848

Galán, B, Mendoza, C G, Calonje, M and Novaes-Ledieu, M: Production, purification, and properties of an endo-1,3-β-glucanase from the basidiomycete *Agaricus bisporus,* Curr Microbiol, 38, 190-193 (1999)

Hirano, T.; Sato, T. & Enei, H. (2004) Isolation of genes specifically expressed in the fruit body of the edible basidiomycete *Lentinula edodes*. Biosci. Biotechnol. Biochem. 68: 468-472.

Hood M.A. (1991) Comparison of four methods for measuring chitinase activity and the application of the 4-MUF assay in aquatic environments. 13: 151-160.

Ikeda, R.; Sugita, T.; Jacobson, E.S. & Shinoda, T. (2002). Laccase and melanization in clinically important Cryptococcus species other than Cryptococcus neoformans. J Clin Microbiol 40, 1214–1218.

Ishida, T.; Fushinobu, S.; Kawai, R.; Kitaoka, M.; Igarashi, K. & Samejima, M: Crystal structure of glycoside hydrolase family 55 β-1,3-glucanase from the basidiomycete *Phanerochaete chrysosporium*, J Biol Chem, 284, 10100-10109 (2009)

Ishibashi, K.; Miura, N.N.; Adachi, Y.; Ohno, N. & Yadomae, T. (2001) Relationship between solubility of grifolan, a fungal 1,3-β-d-glucan, and production of tumor necrosis factor by macrophages in vitro. Biosci Biotechnol Biochem 65:1993–2000

Irie, T.; Sato, T.; Saito, K.; Honda, Y.; Watanabe, T.; Kuwahara, M. & Enei, H (2003) Construction of a homologous selectable marker gene for *Lentinula edodes* transformation. Biosci Biotechnol Biochem.

Jolivet, S.; Arpin, N.; Wichers, H.J. & Pellon, G. (1998). *Agaricus bisporus* browning: a review. Mycol Res 102, 1459–1483.

Kanda, K.; Sato, T.; Suzuki, S.; Ishii, S.; Ejiri, S. & Enei, H. (1996a) Relationships between tyrosinase activity and gill browning during preservation of *Lentinus edodes* fruit-bodies, Biosci Biotechnol Biochem, 60, 479-480

Kanda, K.; Sato, T.; Ishii, S.; Enei, H. & Ejiri, S. (1996b) Purification and properties of tyrosinase isozymes from the gill of *Lentinus edodes* fruiting body, Biosci Biotechnol Biochem, 60, 1273-1278

Kamada, T.; Hamada, Y.; Takemaru T. (1981) Autolysis in vitro of the Stipe Cell Wall in *Coprinus macrorhizus*. 128: 1041-1046

Kaneko, S., and Shishido, K. (2001) Cloning and sequence analysis of the basidiomycete *Lentinus edodes* ribonucleotide reductase small subunit cDNA and expression of a corresponding gene in *L. edodes*. Gene 262: 43-50.

Kim, H., Ahn, J.H.; Görlach, J.M.; Caprari, C.; Scott-Craig, J.S. & Walton, J.D. (2001) Mutational analysis of β-glucanase genes from the plant-pathogenic fungus *Cochliobolus carbonum*. Mol. Plant Microb. Interact 14: 1436-1443

Kitajima, S. & Sato, F. (1999) Plant pathogenesis-related proteins: molecular mechanisms of gene expression and protein function. J Biochem 125: 1-8

Konno, N. & Sakamoto, Y. (2011) An endo-β-1,6-glucanase involved in Lentinula edodes fruiting body autolysis, Appl Microbiol Biotechnol 91:1365-73

Kollár, R.; Reinhold, B.B.; Petráková, E.; Yeh, H.J.; Ashwell, G.; Drgonová, J.; Kapteyn, J.C.; Klis, F.M. & Cabib, E. (1997) Architecture of the yeast cell wall: β(1□□6)-glucan interconnects mannoprotein, β(1–>3)-glucan, and chitin. J Biol Chem 272:17762–17775

Kumar, S.V.S.; Phale, P.S.; Durani, S. & Wangikar, P. (2003) Combined sequence and structure analysis of the fungal laccase family. Bioetchnol Bioengin 83: 386-394

Kuranda, M.J. & Robbins, R.W. (1987) Cloning and heterologous expression of glycosidase genes from *Saccharomyces cerevisiae*. Proc Natl Acad Sci USA 84: 2585-2589

Larriba, G.; Andaluz, E.; Cueva, R. & Basco, R.D. (1995) Molecular biology of yeast exoglucanases. FEMS Microbiol Lett 125: 121-126

Leonowicz, A.; Cho, N.-S.; Luterek, J.; Wilkolazka, A.; Wojtas- Wasilewska, M.; Matuszeska, A.; Hofrichter, M.; Wesenberg, D. & Rogalski, J. (2001). Fungal laccase: properties and activity on lignin. J Basic Microbiol 41, 185–227.

Levitz, S.M.; Nong, S.; Mansour, M.K.; Huang, C.; & Specht, C.A. (2001) Molecular characterization of a mannoprotein with homology to chitin deacetylases that stimulates T cell responses to *Cryptococcus neoformans*. Proc. Natl. Acad. Sci. USA 98: 10422-10427.

Lindsey, R. & Momany, M. (2006) Septin localization across kingdoms: three themes with variations. Curr. Opin. Microbiol. 9: 559-565.

Littlewood, T.D. & Evan, G.I. (1995) Transcription factors 2: helix-loop-helix. Protein. Profile. 2: 621-702.

Mahadevan P.R. & Mahadkar, U.R. (1970) Role of enzymes in growth and morphology of *Neurospora crassa*: cell-wall-bound enzymes and their possible role in branching. J Bacteriol 101:941–947

Martin, F.; Aerts, A.; Ahren, D.; Brun, A.; Danchin, E. G.; Duchaussoy, F.; Gibon, J.; Kohler, A.; Lindquist, E.; Pereda, V.; Salamov, A.; Shapiro, H. J.; Wuyts, J.; Blaudez, D.; Buee, M.; Brokstein, P.; Canback, B.; Cohen, D.; Courty, P. E.; Coutinho, P. M.; Delaruelle, C.; Detter, J.C.; Deveau, A.; DiFazio, S.; Duplessis, S.; Fraissinet-Tachet, L.; Lucic, E.; Frey-Klett, P.; Fourrey, C.; Feussner, I.; Gay, G.; Grimwood, J.;

Hoegger, P. J.; Jain, P.; Kilaru, S.; Labbe, J.; Lin, Y.C.; Legue, V.; Le Tacon, F.; Marmeisse, R.; Melayah, D.; Montanini, B.; Muratet, M.; Nehls, U.; Niculita-Hirzel, H.; Oudot-Le Secq, M. P.; Peter, M.; Quesneville, H.; Rajashekar, B.; Reich, M.; Rouhier, N.; Schmutz, J.; Yin, T.; Chalot, M.; Henrissat, B.; Kues, U.; Lucas, S.; Van de Peer, Y.; Podila, G. K.; Polle, A.; Pukkila, P. J.; Richardson, P. M.; Rouze, P.; Sanders, I. R.; Stajich, J. E.; Tunlid, A.; Tuskan, G.; & Grigoriev, I. V. (2008) The genome of *Laccaria bicolor* provides insights into mycorrhizal symbiosis. Nature 452: 88-92.

Matsumura, H.; Reich, S.; Ito, A.; Saitoh, H.; Kamoun, S.; Winter, P., Kahl, G., Reuter, M., Kruger, D.H. & Terauchi, R. (2003) Gene expression analysis of plant host-pathogen interactions by SuperSAGE. Proc Natl Acad Sci U S A. Dec 23;100(26)

Montero, M.; Sanz, L.; Rey, M.; Monte, E.; & Llobell, A. (2005) BGN16.3, a novel acidic □-1,6-glucanase from mycoparasitic fungus *Trichoderma harzianum* CECT 2413. FEBS J 272:3441−3448

Minamide, T.; Habu, T. & Ogata, K. (1980a) Effect of storage temperature on keeping freshness of mushrooms after harvest. J Jpn Soc Food Sci Technol 27: 281-287

Minamide, T.; Tsuruta, M. & Ogata, K. (1980b) Studies on keeping freshness of Shii-take mushroom (*Lentinus edodes* (Bark) Sing.) after harvest. J Jpn Soc Food Sci Technol 27: 498-504

Minato, K.; Mizuno, M.; Terai, H.; & Tsuchida, H. (1999) Autolysis of lentinan, an antitumor polysaccharide, during storage of *Lentinus edodes*, Shiitake mushroom. J Agric Food Chem 47: 1530-1532

Minato, K.; Kawakami, S.; Nomura, K.; Tsuchida, H. & Mizuno, M. (2004) An exo β-1,3 glucanase synthesized de novo degrades lentinan during storage of *Lentinule edodes* and diminishes immunomodulationg activity of the mushroom. Carbohydr Poly 56: 279-286

Miyazaki, Y.; Nakamura, M.; & Babasaki, K. (2005) Molecular cloning of developmentally specific genes by representational difference analysis during the fruiting body formation in the basidiomycete *Lentinula edodes*. Fungal. Genet. Biol. 42: 493-505.

Miyazaki, Y.; Sakuragi, Y.; Yamazaki, T.; & Shishido, K. (2004) Target genes of the developmental regulator PRIB of the mushroom *Lentinula edodes*. Biosci. Biotechnol. Biochem. 68: 1898-1905.

Moy, M., Li, H.M.; Sullivan, R., White, J.F. Jr & Belanger, F.C. (2002) Endophytic fungal □-1,6-glucanase expression in the infected host grass. Plant Physiol 13:1298−1308.

Muraguchi, H.; Fujita, T.; Kishibe, Y.; Konno, K.; Ueda, N.; Nakahori, K.; Yanagi, S.O. & Kamada, T. (2008) The exp1 gene essential for pileus expansion and autolysis of the inky cap mushroom *Coprinopsis cinerea* (*Coprinus cinereus*) encodes an HMG protein, Fungal Genet Biol, 45, 890-896

Muthukumar, G.; Suhng, S.-H.; Magee, P.T.; Jewell, R.D. & Primerano, D.A. (1993) The *Sccharomyces cerevisiae SPR1* gene encodes a spolulation-specific exo-1,3-β-glucanase which contributes to ascospore thermoresistance. J Bacteriol 175: 386-394

Nanninga, N. (2001) Cytokinesis in prokaryotes and eukaryotes: common principles and different solutions. *Microbiol Mol Biol Rev* 65: 319-333.

Nagai, M.; Kawata, M.; Watanabe, H.; Ogawa, M.; Saito, K.; Takesawa, T.; Kanda, K. & Sato, T. (2003) Important role of fungal intracellular laccase for melanin synthesis: purification and characterization of an intracellular laccase from Lentinula edo- des fruit bodies, Microbiol, 149, 2455-2462

Nagai, M.; Sato, T.; Watanabe, H.; Saito, K.; Kawata, M. & Enei, H. (2002) Purification and characterization of an extracellular laccase from the edible mushroom *Lentinula edodes*, and decolorization of chemically different dyes. Biosci Biotechnol Biochem 73:1042-1047

Nakade, K.; Watanabe, H.; Sakamoto, Y. & Sato, T. (2011) Gene silencing of the Lentinula edodes lcc1 gene by expression of a homologous inverted repeat sequence, Microbiol Res (in press)

Nakazawa, T.; Ando, Y.; Kitaaki, K.; Nakahori, K. & Kamada, T. (2011) Fungal Genet Biol. Efficient gene targeting in ΔCc.ku70 or ΔCc.lig4 mutants of the agaricomycete *Coprinopsis cinerea*. 48(10):939-46.

Namekawa, S.H.; Iwabata, K.; Sugawara, H.; Hamada, F.N.; Koshiyama, A.; Chiku, H.; Kamada, T.; & Sakaguchi, K. (2005) Knockdown of LIM15/DMC1 in the mushroom *Coprinus cinereus* by double-stranded RNA-mediated gene silencing. Microbiology 151: 3669-3678.

Nara, T.; Saka, T.; Sawado, T.; Takase, H.; Ito, Y.; Hotta, Y. & Sakaguchi K (1999) Isolation of a LIM15/DMC1 homolog from the basidiomycete *Coprinus cinereus* and its expression in relation to meiotic chromosome pairing. Mol Gen Genet 262 781-789

Nishikawa, Y.; Tanaka, M.; Shibata, S. & Fukuoka, F. (1970) Polysaccharides of lichens and fungi. IV. Antitumour active *O*-acetylated pustulan-type glucans from the lichens of *Umbilicaria* species. Chem Pharm Bull 18:1431–1434

Oda, K.; Kasahara, S.; Yamagata, Y.; Abe, K. & Nakajima, T. (2002) Cloning and expression of the exo-β-D-1,3-glucanse gene (*exgS*) from *Aspergillus saitoi*. Biosci Biotechnol Biochem 66: 1587-1590

Ooi, V.E.C & Liu, F. (2000) Immunomodulation and anti-cancer activity of polysaccharide-protein complexes. Curr Med Chem 7:715–729

Ohno, N.; Furukawa, M.; Miura, N.N.; Adachi, Y.; Motoi, M. & Yadomae, T. (2001) Antitumor □-glucan from the cultured fruit body of *Agaricus blazei*. Biol Pharm Bull 24:820–828

Oyama, S.; Yamagata, Y.; Abe, K. & Nakajima, T. (2002) Cloning and expression of an endo-1,6-□-d-glucanase gene (neg1) from *Neurospora crassa*. Biosci Biotechnol Biochem 66:1378–1381

Pochanavanich, P.; & Suntornsuk, W. (2002) Fungal chitosan production and its characterization Lett. in Appl. Microbiol. 35: 17-21.

Pirkkala, L.; Nykanen, P.; & Sistonen, L. (2001) Roles of the heat shock transcription factors in regulation of the heat shock response and beyond. FASEB. J. 15: 1118-1131.

Pitson, S.M.; Seviour, R.J.; McDougall, B.M.; Stone, B.A. & Sadek, M. (1996) Purification and characterization of an extracellular (1□□6)-□-glucanase from the filamentous fungus *Acremonium persicinum*. Biochem J 316:841–846

Rogers, G. W.; Jr.; Komar, A.A.; & Merrick, W. C. (2002) eIF4A: the godfather of the DEAD box helicases. Prog. Nucleic Acid Res. Mol. Biol. 72: 307-331.

Rotem, Y.; Yarden, O. & Sztejnberg A (1999) The mycoparasite *Ampelomyces quisqualis* expresses *exgA* encoding and exo-β-1,3-glucanase in culture and during mycoparasitism. Phytopathol 89: 631-638

Saibil, H.R. (2008) Chaperone machines in action. Curr. Opin. Struct. Biol. 18: 35-42.

Sano, H.; Kaneko, S.; Sakamoto, Y.; Sato, T. & Shishido, K. (2009) The basidiomycetous mushroom *Lentinula edodes* white collar-2 homolog PHRB, a partner of putative blue-light photoreceptor PHRA, binds to a specific site in the promoter region of the *L. edodes* tyrosinase gene, Fungal Genet Biol, 46, 333-341

Sakamoto, Y.; Irie, T. & Sato, T. (2005a) Characterization of the Lentinula edodes exg2 gene encoding a lentinan-degrading exo-β- 1,3-glucanase, Curr Genet, 47, 244-252

Sakamoto, Y.; Minato, K.; Nagai, M.; Kawakami, S.; Mizuno, M. & Sato, T (2005b) Isolation and characterization of a fruiting body- specific exo-β-1,3-glucanase-encoding gene, exg1, from *Lentinula edodes*, Curr Genet, 48, 195-203

Sakamoto, Y,; Watanabe, H.; Nagai, M.; Nakade, K.; Takahashi, M. & Sato, T. (2006) *Lentinula edodes* tlg1 encodes a thaumatin-like protein that is involved in lentinan degradation and fruiting body senescence, Plant Physiol, 141, 793-801

Sakamoto, Y.; Nakade, K.; Yano, A.; Nakagawa, Y.; Hirano, T.; Irie, T.; Watanabe, H.; Nagai M. & Sato, T (2008) Heterologous expression of *lcc1* from *Lentinula edodes* in tabacco BY-2 cells results in the production an active, secreted form of fungal lac- case, Appl Microbiol Biotechnol, 79, 971-980

Sakamoto, Y.; Nakade, K.; Sato, T. (2009) Characterization of the post-harvest changes in gene transcription in the gill of the *Lentinula edodes* fruiting body. Curr. Genet. 55, 409-23

Sakamoto, Y.; Nakade, K.; Konno, N (2011) An endo-β-1,3-glucanase, GLU1, from *Lentinula edodes* fruiting body belongs to a new glycoside hydrolase family. Appl Environ Microbiol. 77, 8350-8354

Sato, T.; Kanda, K.; Okawa, K.; Takahashi, M.; Watanabe, H.; Hirano, T.; Yaegashi, K.; Sakamoto, Y. & Uchimiya, H. (2009) The tyrosinase-encoding gene of *Lentinula edodes*, Letyr, is abundantly expressed in the gills of the fruit-body during post-harvest preservation. Biosci Biotechnol Biochem. 73:1042-1047

Shida, M.; Ushioda, Y.; Nakajima, T.; and Matsuda, K. (1981) Structure of the alkali-insoluble skeltetal glucan of *Lentinus edodes*. J. Biochem. 90: 1093-1100.

Seiler S, Plamann M (2003) The genetic basis of cellular morphogenesis in the filamentous fungus *Neurospora crassa*. Mol Biol Cell 14:4352–4364

Spiess, C.; Meyer, A. S.; Reissmann, S.; and Frydman, J. (2004) Mechanism of the eukaryotic chaperonin: protein folding in the chamber of secrets. Trends Cell Biol. 14: 598-604.

Stassen, N.Y.; Logsdon, J.M.; Jr.; Vora, G.J.; Offenberg, H. H.; Palmer, J. D.; and Zolan, M. E. (1997) Isolation and characterization of rad51 orthologs from *Coprinus cinereus* and *Lycopersicon esculentum*, and phylogenetic analysis of eukaryotic recA homologs. Curr. Genet. 31: 144-157.

Suizu, T.; Zhou, G.L.; Oowatari, Y.; and Kawamukai, M. (2008) Analysis of expressed sequence tags (ESTs) from *Lentinula edodes*. Appl. Microbiol. Biotechnol. 79: 461-470.

Tabata, K.; Itoh, W.; Hirata, A.; Sugawara, I.; & Mori, S. (1990) Preparation of polyclonal antibodies to an anti-tumor (1–>3)-β-d-glucan, schizophyllan. Agric Biol Chem 54:1953–1959

Terashima, K.; Yuki, K.; Muraguchi, H.; Akiyama, M. & Kamada, T. (2005) The *dst1* gene involved in mushroom photomorphogenesis of *Coprinus cinereus* encodes a putative photoreceptor for blue light. Genetics 171: 101-108

Thomas, J.O.; & Travers, A.A. (2001) HMG1 and 2, and related 'architectural' DNA-binding proteins. Trends Biochem. Sci. 26: 167-174.

Trudel, J.; Grenier, J.; Potvin, C. & Asselin, A. (1998) Several thaumatin like proteins bind to β-1,3-glucans. Plant Physiol 118: 1431-1438

Turner, E.M. (1974). Phenoloxidase activity in relation to substrate and development stage in the mushroom, *Agaricus bisporus.* Trans Brit Mycol Soc 63, 541–547.

van Gelder, C.W.; Flurkey, W.H. & Wichers, H.J. (1997) Sequence and structural features of plant and fungal tyrosinases. Phytochemistry. 1997 Aug;45(7):1309-23

van de Rhee, M.D.; Mendes, O.; Werten, M.W.; Huizing, H.J. & Mooibroek, H. (1996) Highly efficient homologous integration via tandem exo-β-1, 3-glucanase genes in the common mushroom, *Agaricus bisporus*. Curr Genet 30: 166-173

Vazquez de alanda, C.R.; Correa, J.; San Segundo, P.; Bueno, A.; Nebreda, A.R.; Mendez, E. & del Rey, F. (1991) Nucleotide sequence of the exo-β-1,3-D-glucanase-encoding gene, *EXG1*, of the yeast *Saccharomyces cerevisiae*. Gene 97: 173-182

van Loon, L.C. & van Strien, E.A. (1999) The families of pathogenesis-related proteins, their activities, and comparative analysis PR-1 type proteins. Physiol Mol Pl Pathol 55: 85-97

Wessels, J.G.; & Niederpruem D.J. (1967) Role of a cell-wall glucan-degrading enzyme in mating of *Schizophyllum commune.* J Bacteriol 94:1594–1602

Williams, D.B. (2006) Beyond lectins: the calnexin/calreticulin chaperone system of the endoplasmic reticulum. J. Cell. Sci. 119: 615-623.

Wissmüller, S.; Kosian, T.; Wolf, M.; Finzsch, M.; & Wegner, M. (2006) The high-mobility-group domain of Sox proteins interacts with DNA-binding domains of many transcription factors. Nucleic Acids Res. 34: 1735-1744.

Wagemaker M.J.; Eastwood, D.C.; van der Drift, C., Jetten, M.S.; Burton, K.; Van Griensven, L.J. & Op den Camp, H.J. (2007) Argininosuccinate synthetase and argininosuccinate lyase: two ornithine cycle enzymes from *Agaricus bisporus*. Mycol Res 111: 493-502

Wolf, D.H.; & Hilt, W. (2004) The proteasome: a proteolytic nanomachine of cell regulation and waste disposal. Biochim. Biophys. Acta 1695: 19-31.

Yano, A.; Kikuchi, S.; Nakagawa, Y.; Sakamoto, Y. & Sato, T. (2009) Secretory expression of the non-secretory-type *Lentinula edodes* laccase by *Aspergillus oryzae*. Microbiol Res. 164(6):642-9

Zhao J. & Kwan H.S. (1999) Characterization, molecular cloning, and differential expression analysis of laccase genes from *Lentinula edodes*. Appl Environ Microbiol 65: 4908-4913

7

Feeding Habits of Both Deep-Water Red Shrimps, *Aristaeomorpha foliacea* and *Aristeus antennatus* (Decapoda, Aristeidae) in the Ionian Sea (E. Mediterranean)

Kostas Kapiris
Hellenic Centre for Marine Research,
Institute of Marine Biological Resources,
Greece

1. Introduction

1.1 The deep-water fishes' diet habits: Some notes on the present and the future knowledge

The study of dietary composition and feeding habits of fish has been considered as one of the most important topics in the ecology of animal marine communities. Both the biology of species (as a population level) and their interaction with biological and environmental factors are influenced by the food resources exploited by fish.

At bathyal depths, most of our knowledge on deep-sea demersal communities, mainly for both fishes and large invertebrates - e.g. decapods crustaceans, comes principally from areas where deep-water fisheries are commercially developed. Most deep-sea (>1000 m) fish are active predators, regarded as generalist (i.e. highly diversified diets) and depend mainly on benthopelagic and mesopelagic prey (Mauchline & Gordon, 1986). The distribution of these preys is closely associated with the bottom; however, many demersal fishes feed principally on vertically migrating mesopelagic organisms such as myctophids and cephalopods. Crustacean zooplanktivores constitute the majority of deep-sea pelagic fish species and families examined. Less common are predators that primarily ingest soft-bodied or gelatinous zooplankton, gastropod molluscs and polychaete worms. These categories of predators are generally represented by a few individual species within different families (Randall & Farrell, 1997)

Many deep-sea fish show diel and seasonal feeding patterns (e.g. Macpherson, 1980; Mauchline & Gordon, 1984; Atkinson, 1995), which have been often related to the temporal vertical migrations of their mesopelagic prey (Gartner et al., 1997). To understand temporal (e.g. daily, seasonal) and spatial changes in the diet and trophic habits of bathyal fish it is crucial simultaneous sampling of their trophic resources (Madurell, 2003). However despite the accepted role of this fauna in the bathyal food webs, little is known about their interactions and their dynamics.

To date, studies on the feeding habits of deep-sea fish have focused mainly on depth related changes and only few studies have addressed aspects of seasonal or diel feeding cycles. Studies on food resource partitioning are scarce in non-littoral marine environments, with some studies performed in deep-sea communities at a community level and within a single feeding guild (e.g. Carrassón & Cartes, 2002). In general, food is considered as the major limiting factor in the functioning of deep-water ecosystems and trophic aspects have been considered as the most important factor on deep-sea faunal community organization (Jumars & Gallagher, 1982).

The knowledge concerning the deep-sea organisms' biology remains almost in its infancy. Many aspects about their diet are still speculative. Many lags exist concerning the type of the bottom, the spatial and temporal distribution of many fishes and the taxonomic composition of the fauna. The result of further detailed studies could offer important elements to our learning about the taxonomic composition, life history and physiology studies, such as feeding habits (Randall & Farrell, 1997).

1.2 Feeding habits in Decapod crustaceans of shallow and deep-sea waters – General aspects

The much diversified diet of penaeids (shrimps or prawns) was first described by Williams (1955) who studied the stomach content of prawns from the eastern United States. Generally speaking, the condition of the gut contents prevents the use of the most methods of quantifying diet. Techniques such as weighting the food, measuring the size of prey or reconstituting prey cannot be used (Dall et al., 1990). Decapods can catch mobile free-swimming prey; the remains of fish and squid form a major part of the diet of several species and their preferable animal food is other small crustaceans, polychaets, mollusks (Dall et al., 1990), plant remains (Kuttyamma, 1974) and a very small portion of bacteria (Moriarty & Barclay, 1981).

Most shrimps spend the day buried in the substratum and emerge and feed at night. Because of their small stomach, they feed several times each night in order to obtain sufficient food. Under natural conditions, where food items are small and dispersed, searching and ingestion probably continue through the night and rates of ingestion and ejection are probably similar (Dall, 1968). This allows the foregut to be filled repeatedly and enables the shrimps to take in considerably more food if they fed only once per night (Dall et al, 1990).

Decapods can even be dominant in some deep-water regions, including the deep Mediterranean, where subtropical species dominate in terms of biomass (Cartes & Sardà, 1992). It has been suggested that this dominance is due to the low metabolic rates of this taxon, which has low feeding rates and hence is better adapted to live under oligotrophic conditions than, for instance, fishes (Cartes & Sardà, 1992). In bathyal ecosystems, decapods occupy a variety of ecological niches and exhibit a similar range of trophic levels to fish (Polunin et al., 2001). They also have a wide variety of feeding habits or guilds (Cartes et al., 2002), ranging from deposit feeders to carnivores, the latter including specialized species preying on benthos (e.g. Crangonidae: Lagardère, 1977) or macrozooplankton (e.g. Pandalidae: Cartes, 1993a). As a consequence, decapods are an ideal group in which to analyze changes in the structure and dynamics of ecosystems exposed to temporal (e.g. Seasonal) or spatial (e.g. depth) gradients (Cartes et al., 2007).

2. Mediterranean deep sea waters

The Mediterranean Sea – *Mare Nostrum, Mare Internum* or *Mediterraneum* is a relatively small, deep basin that is surrounded by continents. It is 4000 km in length and is located between 30° and 46° N and 5.50° W and 36° E (excluding the Black Sea). The Mediterranean Basin is subdivided by a series of tranverse ridges with a north-south trend. These ridges exist in the western (Alboran, Balearic, Tyrrhenian basins) and the eastern part (Ionian, Aegean, Levantine) (Sardà et al., 2004 and references herein).

Deep-sea ecosystems include the waters and sediments beneath approximately 200 m depth (Emig & Geistdoerfer, 2004) and represent the world's largest biome, covering more than 65% of the earth's surface and including more than 95% of the global biosphere. The deep-sea domain of the Mediterranean is divided in three zones: bathyal, abyssal and hadal (Table 1). The edge of the continental shelf is the boundary between the neritic (inshore) domain and the deep-sea oceanic (offshore) domain. Its depth varies with the ocean or Sea.

The deepest point in the Mediterranean, 5,121 m, is found at the North Matapan-Vavilov Trench, Ionian Sea. The deep-sea floor includes regions characterized by complex sedimentological and structural features: (a) continental slopes, (b) submarine canyons, (c) base-of-slope deposits, and (d) bathyal or basin plains with abundant deposits of hemipelagic and turbidity mud's (Danovaro et al., 2010).

	Neritic Domain	Deep-Sea Oceanic Domain		
	Continental Shelf	Bathyal Zone	Abyssal Zone	Hadal Zone
Limits in m				
World Ocean	0 m to Shelf-break	Shelf-break to 3,000 m	3,000 to 6,000 m	More than 6,000 m
Mediterranean Sea	0 to 100-110 m	100-110 to 3,000 m	3,000 to 5,093 m	-
Extension in % of the entire area				
World Ocean	7 %	15 %	77 %	1 %
Mediterranean Sea	15 %	72 %	13 %	-

Table 1. Comparison of the ranges in depth of the deep-sea zones of the World Ocean and the Mediterranean Sea (Emig & Geistdoerfer, 2004).

A further unique feature of the Mediterranean is that it is one of the few warm deep-sea basins in the world, where temperatures remain largely uniform at around 12,5-14,5°C at all depths, with high salinity (38,4-39,0 PSU) and high oxygen levels (4,5-5,0 ml, Hopkins, 1985). The constant temperature and salinity regime of the Mediterranean contrasts with the Atlantic at comparable latitudes, where temperature decreases and salinity increases with depth.

The Eastern and Western Mediterranean display important geological and biological differences. First, the Eastern Mediterranean is geologically more active, because it is a contact zone between 3 major tectonic plates: African, Eurasian and Arabian. The Western Mediterranean is relatively featureless in comparison, although still not devoid of unique environments. Biologically, the Western Mediterranean, whilst still oligotrophic by North Atlantic standards, has relatively high primary production, especially in the Gulf of Lions, due to the river Rhone runoff and wind mixing. The Eastern Mediterranean has very low primary production. The existing deep Mediterranean appears to be much younger than any other of the world's deep ocean and only a small fraction of specialized taxa exists in its deep-sea fauna. Traditionally the Mediterranean Sea is one of the most intensively investigated areas of the world in both terrestrial and coastal marine biodiversity, but it lags other regions of the world in studies of its deep-sea fauna.

2.1 Mediterranean deep sea fauna

The Mediterranean appears to be a remarkable natural laboratory for the study of the processes of recent colonization as related to the unique history of each of the two great Mediterranean basins (western and eastern) (Emig & Geistdoerfer, 2004) and, due to the presence of few endemic species, is a important centre of evolution. Its fauna is composed mainly of primitive taxonomic groups among the phyla represented, whereas a small fraction of specialized taxa exists in the deep-sea fauna. Despite its small dimensions (0,82% of the ocean surface), the basin hosts more than 7,5% of global biodiversity (Bianchi & Morri, 2000). However, this information is almost completely confined to coastal ecosystems, and data on deep-sea assemblages are still limited (e.g. Ramirez-Llodra et al., 2009).

The stable environment of the deep Mediterranean Sea permits to the biotic (e.g. trophic) factors may have a comparatively strong influence on the ecology and biology (e.g. food intake and reproduction) of deep-Mediterranean species (Madurell & Cartes, 2006). Also, the depth overlap between and the depth range inhabited by megafaunal fish and decapod crustaceans (Cartes and Carrasson, 2004) are mainly explained by trophic variables (e.g. trophic level).

The deep Mediterranean fauna displays a number of characteristics that differentiate it from other deep-sea faunas of the world's oceans (Bouchet & Taviani, 1992): *i)* the high degree of eurybathic species; *ii)* absence (or low representation) of typical deep-water groups, such as macroscopic foraminifera (Xenophyophora), glass sponges (Hexactinellida), Sea-cucumbers of the order Elasipodida, primitive stalked Sea-lilies (Crinoidea) and tunicates (Sea-squirts) of the class Sorberacea (Monniot & Monniot, 1990); and *iii)* the number of endemic species (26.6% of the Mediterranean fauna: Ruffo, 1998) declines with increasing depth, with comparatively low endemisms below 500 m (see also Fredj & Laubier, 1985).

The existed quantitative data from this basin are scarce. Several investigations have described low-abundance and low-diversity conditions of marine invertebrates in the Eastern Mediterranean (Tselepides & Eleftheriou, 1992). Scientific knowledge of deep megafaunal communities (mainly fish, crustaceans and cephalopods) was limited to the bathymetric range exploited by fishing (down to 800-1000-m) until the early 1980's, when scientific expeditions began quantitatively sampling the bathyal grounds in the Mediterranean. Such studies in the Western and Central Mediterranean have focused on the two most abundant groups below 600 m depth: fishes (D'Onghia et al., 2004) and decapod crustaceans (e.g. Company et al., 2004). At depths below the 1500 m, there is an increase in the relative abundance of crustaceans in comparison to fish (Company et al., 2004). This change in the relative abundance of these two groups has been explained by the low food availability at greater depths and the higher adaptation of crustaceans to low energy levels (e.g. Company et al., 2004).

2.2 Mediterranean deep sea crustacean fauna

Decapod crustaceans are one of the dominant megafaunal groups in the deep-sea communities of the Mediterranean Sea (Sardà et al., 1994a). The relatively oligotrophic nature of Mediterranean waters has been presented as one of the environmental factors contributing to the high abundance of decapod crustaceans in comparison with other oceans, in which other megafaunal invertebrates, chiefly echinoderms, predominate (Tyler & Zibrowius, 1992). However, though the Mediterranean is a relatively small Sea compared with other oceans, existing data and our understanding of the continental margins at depths below 2000 m lags behind. Decapods are much more abundant than other invertebrate groups in the Mediterranean, in contrast to more productive oceans like the Atlantic, where echinoderms are the dominant invertebrate group (e.g. Sardà et al., 1994a), because these crustaceans be more competitive than other invertebrate or vertebrate megafauna in oligotrophic environments (Maynou & Cartes, 2000).

In the W. Mediterranean 28 decapod species were identified, including 6 Dendrobranchiata, 1 Stenopodidean, 7 Caridea, 2 Thalassinidea, 2 Palinura, 3 Anomura, and 7 Brachyura.. The most pronounced qualitative changes in the fauna were recorded between 1000 and 1200 m and at around 2000 m. The bathyal decapod fauna mainly composed by species belonging to the families Crangonidae, Galatheidae and Geryonidae, and the genera *Nematocarcinus* and *Stereomastis.* In addition to this the tropical species *A. antennatus, Acanthephyra eximia,* and *Plesionika acanthonotus* are widely distributed and frequent in the deep western Mediterranean (Cartes, 1993b). Thirty nine decapod species have been reported in the Eastern Ionian Sea (E. Mediterranean), of which eight were Dendrobranchiata and 31 Pleocyemata (17 Caridea, 9 Brachyura, 3 Anomura, 1 Astacidea and 1 Palinura) (Politou et al., 2005). Concerning their depth distribution, 30 species were found in the depth zone 300-500 m, with *Parapenaeus longirostris* being the most abundant species. Of the 27 species caught in the zone 500-700 m, *A. foliacea* and *Plesionika martia* were the most abundant. In the zone 700-900 m, 19 species were found and *A. foliacea* with *A. antennatus* were the most numerous. Finally, the 18 decapod species encountered in the zone 900-1200 m showed low abundance, and *Sergia robusta* with *Polycheles typhlops* predominated in numbers. From the identified decapods, *Acanthephyra eximia, Philoceras echinulatus* and *Pontophilus norvegicus* were mentioned for the first time in the E. Ionian Sea. Some other species, such as *Acanthephyra pelagica, Geryon longipes, Munida tenuimana, Paromola cuvieri, Parthenope macrochelos, Pasiphaea multidentata, Plesionika narval, Polycheles typhlops, Sergestes arachnipodus* and *Sergestes arcticus* have been reported for the area only in the gray literature.

3. Aristeidae: Distribution and particular hydrological conditions

In updated systematic approaches (Pérez-Farfante & Kensley, 1997) both deep-sea red shrimps - *A. antennatus* and *A. foliacea* - are the only Mediterranean representatives of the Aristeidae family (superfamily Penaeoidea, Order: Decapoda, Sub-order: Dendrobranchiata). The species of the family (*Aristaeomorpha, Aristeus* and *Plesiopenaeus*), all large-size commercial shrimps, occur in deep water off the continental shelf. Important morphological characteristics of both species are i) light exoskeletons and long pleopods suggesting good swimming ability; ii) secondary sexual dimorphism concerning body size and the rostrum; and iii) an open thelicum. The spermatophores are larger in *A. antennatus,* in relation to a greater fecundity: in fact the females of *A. antennatus* produce about four times more eggs than *A. foliacea* females of the same size (Orsi Relini & Semeria, 1983). The life history of the two species therefore begins with a very different energy budget and probably body development.

3.1 *Aristaeomorpha foliacea*: Distribution and importance

The giant red shrimp or deep-sea red shrimp *A. foliacea* (Risso, 1827) is a species of a very wide geographical distribution in the world. It occurs in the Mediterranean Sea and the eastern Atlantic, the western Atlantic, the Indian Ocean and the western Pacific from Japan to Australia, New Zealand and the Fiji Islands (Pérez Farfante & Kensley, 1997) Gracia et al. (2010) recently explored deep waters off the Yucatan Peninsula in Mexico and showed that *A. foliacea* represents a potential fishing resource. Nowadays, *A. foliacea* constitute a valuable deep shrimp fishery off the south-eastern and southern sectors of the Brazilian coast (Dallagnolo et al., 2009). The giant red shrimp has been recently found in large quantities in the Colombian Caribbean Sea (Paramo & Urlich, in press) (Figure 3).

In the Mediterranean, the species is of great economic interest and, together with *A. antennatus* (Risso, 1816), represents the main target species of the slope trawl fisheries down to 800-1000 m (Demestre, 1994; Ragonese et al., 1994a,b; Sardà & Cartes, 1994b; Matarrese et al.,1997). This species is heavily exploited in Western Mediterranean, and is currently fished in the Central Mediterranean; its stocks are pristine in the Eastern Mediterranean (Bianchini & Ragonese, 1994; Papaconstantinou & Kapiris, 2003; Gönülal et al., 2010) and its exploitation is not yet been developed.

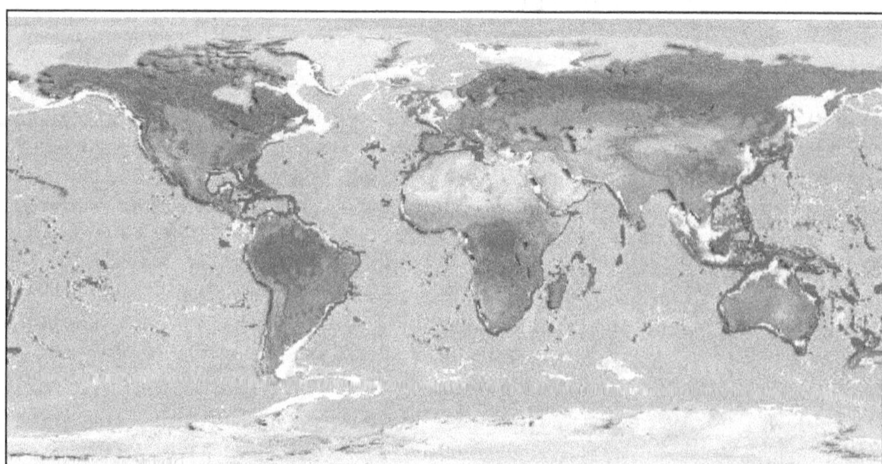

Fig. 2. Geographical distribution of *A. foliacea* (Source: http://www.aquamaps.org/receive.php).

The economic importance of the giant red shrimp in the Mediterranean enhanced the scientific interest of the study and evaluation of its stocks. Most scientific information on *A. foliacea* comes from the central Mediterranean, where the species is relatively abundant and exploited by the commercial fishery (e.g. D'Onghia et al., 1998). It concerns its biology (e.g. D'Onghia et al., 1994; Levi & Vacchi, 1988; Mura et al., 1997), ecology (e.g. Ragonese et al., 1994a) and fisheries (e.g. Ragonese, 1995; Matarrese et al., 1995). In the eastern Mediterranean (Greek waters), knowledge of the species was recently obtained. Some information on its distribution (Kallianiotis et al., 2000; Kapiris et al., 2001c), morphometry (Kapiris et al., 2002; Kapiris, 2005) and biology (Kapiris et al., 1999 ; Kapiris & Thessalou-Legaki, 2001b, 2006, 2009; Papaconstantinou & Kapiris, 2003), feeding (Kapiris et al.,2010) and fishery (Mytilineou et al.,2006) appeared in the literature. According to the present state of knowledge the species depth distribution ranges between 123 and 1047 m, with a maximum abundance from 400 to 800 m in most areas. The maximum depth of occurrence was found to be 1100 m for the whole Mediterranean basin (Politou et al., 2004).

Fig. 3. Specimens of *A. foliacea* (Source: http://www.google.gr).

3.2 *Aristeus antennatus*: Distribution and importance

During the last twenty years a variety of aspects of the blue-red deep water shrimp (*A. antennatus* Risso, 1816) (Figure 4) have been studied in detail in the western, central and eastern Mediterranean Sea, such as fisheries (e.g. Demestre & Martín, 1993; Bianchini & Ragonese, 1994; Sardà et al., 1998; 2003; Papaconstantinou & Kapiris, 2001; D'Onghia et al., 2005), biology (e.g. Kapiris & Thessalou-Legaki, 2001a, 2006, 2009; Kapiris et al., 2002; Kapiris & Kavvadas, 2009; Matarrese et al., 1997; Sardà & Cartes, 1993; Mura et al., 1998; Follesa et al., 1998; Orsi Relini & Relini, 1998; Sardà et al.,1998), ecology (e.g. Sardà & Cartes, 1997; Cartes & Maynou, 1998; Kapiris et al., 1999; Kapiris & Thessalou-Legaki, 2011), and physiology (e.g. Company & Sardà, 1998; Puig et al., 2001).

In the Mediterranean, this important commercially and biologically species may be fished from depths of 80 m along the Algerian coast at night (Nouar, 2001), with more abundant distribution between 400 m and 800 m in Tyrrhenian waters (Aquastudio, 1996) (Figure 5). Its eurybathic distribution ranges from 100 and 150 m to nearly 1000 m in the western Ionian Sea (south Italy, Relini et al., 2000), down to 800 in the eastern Ionian (Papaconstantinou & Kapiris, 2001) and 900-1000 m off Catalonia (Demestre & Martín, 1993; Sardà et al., 1998).

Nevertheless, experimental catches (Sardà et al., 2004) have been made down to a depth of 3300 m. This broad depth distribution range for this species has led to a number of hypotheses concerning its ecology and possible relationships between the exploited populations on the upper and middle slope and the non-exploited populations dwelling deeper on the lower slope (Sardà et al., 2003). Its biology (reproduction, sex-ratio, feeding habits, and population dynamics and fisheries) is relatively well known down to 800 m, where fishery occurs.

Fig. 4. Female and male individual of *A. antennatus* (Source: http://cobmedits2011.wordpress.com/produccion-scientifica/evaluaciones)

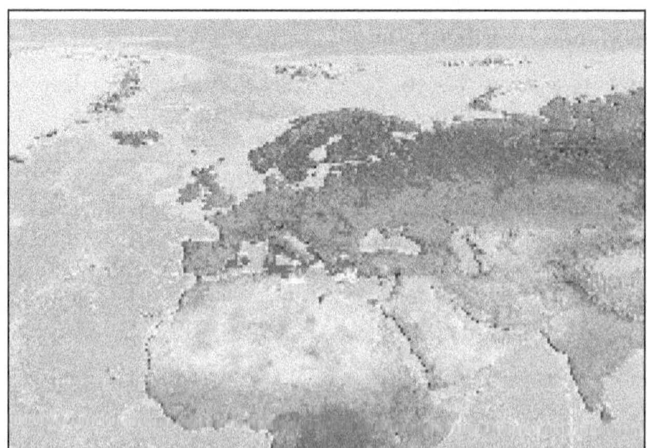

Fig. 5. Geographical distribution of *A. antennatus* (Source: http://fishbase.sinica.edu.tw/slp/SpeciesSummary.php?group=All&ID=ITS-96062).

An increasing abundance gradient from the western to the eastern Mediterranean is confirmed by several previous works for these both deep-sea red shrimps. The abundance of *A. foliacea* increases gradually eastwards, from the Tyrrhenian Sea to the Straits of Sicily and the waters around Greece, where it becomes more abundant than *A. antennatus* (Politou et al., 2003). Different hydrological conditions (i.e. temperature and salinity) between the westernmost and the easternmost areas have been reported to affect the species distribution (Relini & Orsi Relini, 1987). *A. foliacea* is considered to be linked to warmer and more saline water masses than the other deep Sea shrimp *A. antennatus* (Ghidalia & Bourgeois, 1961). Furthermore, the eastern Mediterranean deep water transient (Klein et al., 1999) may play a role in the increased abundance of *A. foliacea* in the eastern Ionian Sea, since this event is associated with a significant upward nutrient transport, which is most pronounced in the eastern Ionian Sea, and may result in greater biological productivity.

4. Diet studies on Aristeids mainly in the E. Mediterranean – The aim of this study

Both aristeids present an increased diversity in their diet (Burukovsky, 1972; Lagardère, 1977; Relini Orsi & Wurtz, 1977; Cartes, 1994, 1995; Gristina et al., 1992; Maurin & Carries, 1968; Chartosia et al., 2005).). Brian (1931), for the first time, has studied the alimentary habits of *A. foliacea* and *A. antennatus* in the Ligurian Sea and stressed the big diversity of prey types consumed by the two shrimps (e.g. pelagic, benthic and benthopelagic organisms). In the Ionian Eastern Sea the diet and the feeding habits of both aristeids have been studied in details (Kapiris & Thessalou-Legaki, 2011; Kapiris et al., 2010).

The object of this chapter is to provide a detailed description of the feeding habits of both deep-water red shrimps in the Eastern Ionian, in relation to Season, size and sex. New information concerning its feeding patterns provides greater insight concerning the population ecology of these important and unexploited resources of the Greek Seas. In addition, such data could serve in the comparison of its life history traits along the Mediterranean.

4.1 The study area

Few studies have been carried out about the dominant demersal fish species found on the upper-middle slope (between 473 and 603 m) in the Ionian Sea (e.g. D'Onghia et al., 1998; Kallianiotis et al., 2000; Labropoulou & Papaconstantinou, 2000; Madurell et al., 2004). The results of those studies indicated that the Ionian demersal ichthyofauna is similar to the other eastern Mediterranean areas. Dominant species in the eastern Mediterranean, such as *H. mediterraneus*, and *C. agassizi* which are plankton feeders, are rare (or absent) in the Catalan Sea (Stefanescu et al., 1994). Macrofauna from the eastern Mediterranean decrease in biomass below 400m and there is a significantly lower biomass of meiofauna than in the Western Basin (Tselepides & Eleftheriou, 1992; Danovaro et al., 1999). These low levels of benthos biomass may reinforce the dominance of top predators feeding on planktonic resources in the Ionian Sea.

Exploratory sampling of *A. foliacea* and *Aristeus antennatus* took place along the south coast of the Greek Ionian Sea, between Zakinthos Island and Peloponnisos Peninsula (Figure. 6). A total of 92 hauls were taken during 12 experimental trawl survey cruises on a monthly basis (December 1996–November 1997). Samples were collected by the commercial trawler Panagia Faneromeni II (26 m in length, 450 HP) using a net with a cod-end mesh size of 18 mm from knot to knot. The results of the feeding habits and diet of both aristeids in the Eastern Ionian are given below.

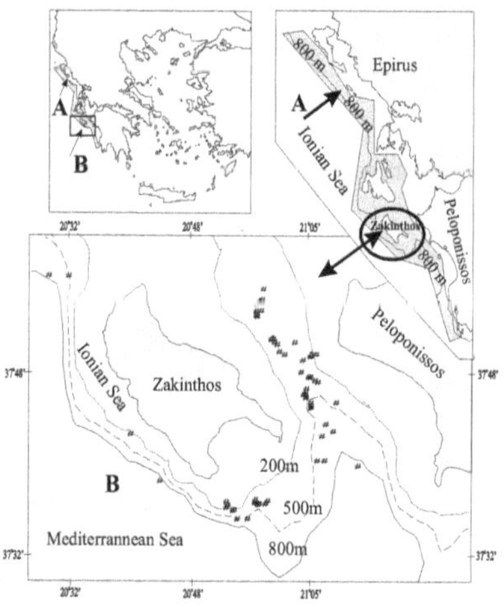

Fig. 6. Study area in the Eastern Greek Ionian Sea

4.2 *foliacea's* feeding habits

4.2.1 Feeding activity and food quality

The highly diversified diet observed in *A. foliacea* is typical of bathyal penaeoideans in the Western Mediterranean (Cartes, 1995). The feeding activity of *A. foliacea* in the Eastern Ionian Sea was examined studing the stomach fullness according the equations (i) wet food weight (g) per 100 g shrimp wet weight [% body weight (BW) Wet = (SWW/BW) x 100] and (ii) dry food weight (g) per 100 g wet weight [%BWDry = (SWD/BW) * 100] (Héroux & Magnan, 1996). The nutritional quality (food quality) of the preys has been estimated by two ways: (a) % dry weight (DW) = (SWD/SWW) 100 and (b) % ash free dry weight (AFDW) = (AFDW/SWD) x 100, where SWW=stomach wet weight, g, SWD= stomach dry weight, gr after 24 h of oven drying at 70° C), ash-free dry weight (AFDW; as loss on ignition at 450° C for 3 h) and BW is the body weight. All the weights were measured to an accuracy of 0.0001 g). The food quality indices are a measure of total organic matter and form a better estimation of food value that wet weight, which includes substantial amounts of inorganic material (Hiller-Adams & Childress, 1983). The stomach fullness of *A. foliacea* varied Seasonally in both sexes and both fullness indices (%BW Wet, %BW Dry) were significantly higher in females than in males for each Season. The maximum values of %BW Wet in both sexes occurred in winter and the minimum in spring.

Diet quality (%BW Dry and %AFDW) also differed significantly among Seasons for both sexes of *A. foliacea*. In general, few significant differences in food quality were detected between males and females for each Season. Males and females of *A. foliacea* presented the highest values of both quality indices in spring and the minimum in winter.

The feeding habits of this decapods identified in the Ionian Sea are generally comparable to those reported in other regions of the Mediterranean, such as in the Catalan Sea (Cartes, 1995), Sicilian Channel (Gristina et al., 1992) and Aegean Sea (Chartosia et al., 2005). Any difference observed in the whole Mediterranean, such as food diversity, different food categories and mean number of prey could be due to bottom morphology (Cartes 1995) and to the oligotrophic conditions of the Eastern Mediterranean. This characteristic of the Eastern Mediterranean could also explain the increased number of pelagic prey consumed by *A. foliacea* compared to the western part of the basin (Cartes, 1995). The considerably higher water temperature of the Eastern Mediterranean (Politou et al., 2004) may also play a role, resulting in a higher metabolic rate of this species, in comparison with those from the western part of the basin.

Trophic diversity ((H', Shannon-Wiener index) varied slightly among Seasons in both sexes (Figure 7) and no statistically significant differences were established between sexes. The maximum diversity (3.00 and 3.04 for males and females, respectively) and mean number of prey items (2.9 and 3.1 for males and females, respectively) were found in summer for both sexes of *A. foliacea*.

The observed low number of empty stomachs [(number of empty stomachs per number of stomachs examined) * 100] (Hyslop, 1980) in the present study, ranging from 4,5 to 18,1%, indicating either a high feeding rate or slow digestion rate, could be explained by their high metabolic rates. The lowest proportion of empty stomachs of *A. foliacea* was found in spring for both sexes, followed by summer. In contrast, the highest number of empty stomachs was found in autumn for females and summer for males.

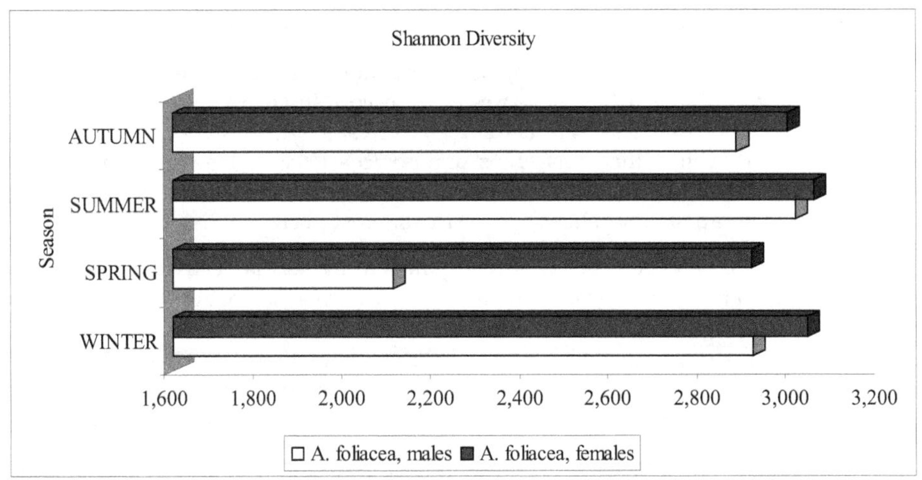

Fig. 7. Diversity index (H', Shannon-Wiener index) values for *A. foliacea* per sex and season in the Ionian Sea.

In general, a decrease in diversity and mean prey items with increasing overlap was observed. In the Eastern Ionian Sea, the giant red shrimp fed on a greater proportion of pelagic resources and prey with a good swimming ability, such as the natantian decapods, and to a lesser extent on benthic prey, indicating that this shrimp is an active and effective predator of the bathyal zone in the Eastern Mediterranean. The characteristic of its active predation could be also confirmed by the very low abundance of infaunal and epibentic prey (e.g. polychaetes, bivalves and gastropods) in the stomachs of this species. The increased abundance of fishes and cephalopods in their foreguts most probably reflects the great scavenging ability of this species. In any case, this does not exclude the possibility that this species feeds actively upon fishes and cephalopods.

4.2.2 Food habits in relation to sex, season and size

The diets of both sexes of *A. foliacea* consisted of 60 different prey categories (most as species-level prey categories). The preys belonged chiefly to three major groups: (i) crustaceans particularly decapods, reptantia (anomurans, brachyurans), amphipods, euphausiids, ostracods, copepods, mysids, tanaidaceans, cumaceans, (ii) cephalopods and (iii) fishes. These three prey categories constituted 72–82% of the relative abundance and total occurrence for males and 70–88% of the relative abundance and the total occurrence in females. The most dominant natantians found were the nektobenthic *Plesionika martia, Plesionika heterocarpus* and *Plesionika giglioli*, followed by *Pasiphaea sp., Sergestes sp.* and *Solenocera sp.* Some appendages from *Aristeus antennatus* were also found mainly in female *A. foliacea*. These findings could be accidental, as they were found in the sampling stations where both species coexisted and, thus, some body appendages could have been destroyed and mixed during the net tow (net feeding). It is also possible that the smaller individuals of each species were consumed by larger adults of the other, due to their voracious character, but further study of this hypothesis is required. Among cephalopods, the dominant species were *Abraliopsis pfefferi, Pyroteuthis margarifera* and *Abralia veranyi*. For fishes, specimens of Myctophidae and Macrouridae were the most abundant in the foreguts.

Only a partial differentiation in the feeding behaviour, in terms of both diet composition and feeding activity, was observed between sexes of *A. foliacea*. In general, both sexes fed upon natantian decapods, particularly *Plesionika spp., Sergestes sp., Pasiphaea sp.*, and fishes throughout the year, while 'other crustaceans' and polychaetes were ingested on a secondary basis. The consumption of the same prey items, but in different abundance and occurrence, may be attributed to sexual dimorphism and to size difference between the sexes.

In general, the existence of regular Seasonal rhythms in the feeding activity of deep water species is related mainly to Seasonal fluctuations in various factors including the abundance of their prey, depth, local geographical characteristics, submarine canyons, bottom type, Seabed features, Seasonal horizontal or diurnal vertical migrations, etc. (Cartes 1993, 1998). In the Eastern Ionian Sea the Seasonal feeding habits of the giant red shrimp seem to be related to reproduction, and perhaps to other biological processes, and food availability.

High observed values of trophic overlap between Seasons for both sexes indicated that Season is not the main factor affecting the diet of deep-water shrimps in the Eastern Ionian Sea. In spite of this, most feeding activity values (empty stomachs, quality indices, mean number of prey items found into the stomachs, diversity index) support the finding that feeding activity increased during spring–summer for both sexes. This increase could be attributed to the increased reproductive activity (gonad maturity, egg-laying) observed in this period (Papaconstantinou & Kapiris 2001, 2003). In addition, copulation begins at the end of winter and by spring almost all females are inseminated (Kapiris, 2004). The minimum value of the stomach fullness in spring, in combination to the highest food quality value and the lowest vacuity index in females in the same Season, suggests that egg maturation is connected to the feeding habits of *A. foliacea*. During winter, *A. foliacea* had the highest stomach fullness, but with decreased food quality. This increase of food consumption by the giant red shrimp of the Ionian Sea during the pre-reproductive period has also been observed in *A. antennatus* off the Balearic Islands. Increased feeding rates could be the main reason for its egg development and could allow earlier gonad maturity (Cartes et al., 2008a).

Besides the Seasonal feeding adaptation to the biological requirements (reproductive process), food availability also plays an important role for these species in the Eastern Ionian Sea. The highest densities in the suprabenthic fauna (mysids, cumaceans, amphipods, isopods, tanaidaceans) have been observed during spring, but zooplankton (chiefly copepods, ostracods and chaetognaths) were more abundant in summer and autumn. Such fluctuations in food availability have also been shown in the diets of both sexes of *A. foliacea* in this study. Thus, the diet of the giant red shrimp probably reflects localized forage assemblages rather than a preference for specific items.

The size-related changes in diet composition are an important factor in determining ecological relationships of marine organisms during their life span. Comparison of diet composition, dietary diversity, and feeding activity among small, medium and (only for females) large individuals reveals that this decapod undergoes slight changes in feeding habits with increasing body size, as well as gonad maturity, in the Eastern Ionian Sea. Small males and females (immature individuals) consumed fewer prey due to their smaller stomachs, with more frequent occurrence of epibenthic prey in their foreguts. Larger, mature individuals of both sexes are more efficient predators due to their greater swimming ability and larger mandibles. A positive trend of ingesting larger prey with increased size was observed only for females. This is the first time where this gradation, probably due to the population structure and to morphological variation among size classes and sexes, has been observed for *A. foliacea*. In general, somatic growth and gonad development induce a change in this species' feeding behavior as the body grows an increase in the mean weight of prey and a decrease in the mean number of prey items per stomach was obvious. However, almost the same prey occurred in the stomachs of small, medium and large specimens, but in different proportions

4.3 Aristeus antennatus' feeding habits

4.3.1 Feeding activity and food quality

A differentiation has been presented in *A. antennatus* diet according to the depth in the western Mediterranean (Cartes, 1994), the feeding time (Cartes, 1993a) and the daily consumption of food (Maynou & Cartes, 1997, 1998; Cartes & Maynou, 1998). The diet of *A. antennatus* changed as a function of depth at around 1000 m depth in the Catalan Sea, as a function of Seasonality influences by planktonic prey in deeper zones and by possible nocturnal movements upward along the slope canyons (Cartes, 1993a, 1994; Cartes et al., 2010). The importance of spatial patterns in its diet and feeding habits and the main environmental variables controlling these trophic aspects has been studied by Cartes et al.(2008b) in Western Mediterranean. In the whole E. Mediterranean, the feeding habits of *A. antennatus* have been studied in detail in the Ionian (Kapiris & Thessalou-Legaki, 2011) and the Aegean Sea (Chartosia et al., 2005).

The observed low number of empty stomachs in the Greek Ionian (mean value of the empty stomachs in males was 6,53 and for females was 8,54) could be explained by their high metabolic rates (Company, 1995). Significant statistical differences amongst the Seasonal medians of both fullness indices were found [%BW Wet (for both sexes) and %BW Dry (only in females)]. The maximum values of %BW Wet were determined in winter in both sexes and the minimum in spring. Both fullness indices were statistically higher in females than those of males (Figure 8).

Significant statistical differences amongst the Seasonal medians of both indices of food quality (%DW, %AFDW) were established only for females. Females presented a lower value of %DW and higher of %AFDW than males, in spring, while their highest values of both quality indices were found in spring (Figure 8).

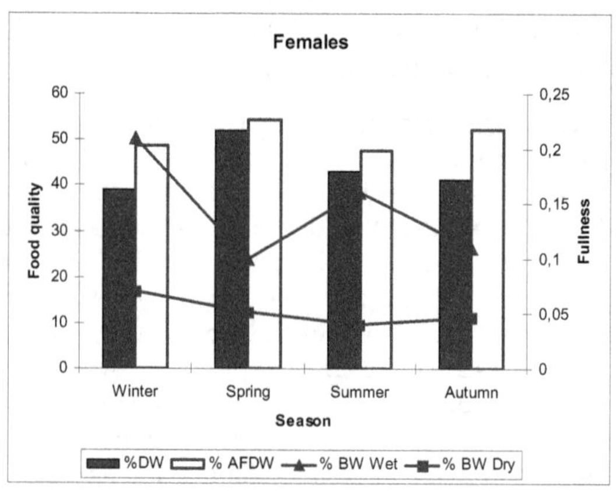

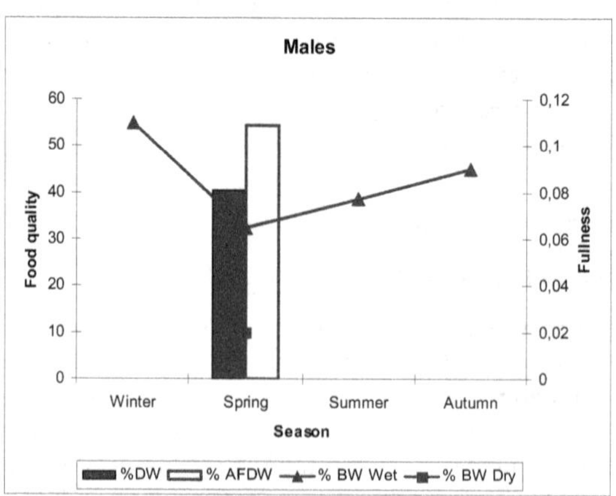

Fig. 8. Seasonal values of stomach fullness and food quality of both sexes of *A. antennatus* in the Greek Ionian Sea.

The diet of *A. antennatus* both sexes consisted of 54 prey categories. These prey items belonged mainly to smaller crustaceans (e.g. natantian decapods, *Plesionika sp., Sergestes sp.*, euphausiids, tanaidaceans), molluscs primarily gastropods, bivalves, polychaetes (Eunicidae, Spionidae, and Nereididae), chaetognaths and, to a lesser extent, fishes. The above prey categories consisted of 71–82% of the relative abundance and total occurrence for males and 61–81% of the relative abundance and the total occurrence in females. Its diversified diet in the present study area consists of increased endobenthic and epibenthic invertebrates and includes organisms that are related with the Seabed, nekton and decapods. This species is among the few megabenthic predators whose diet is mainly based on benthos in the deep Mediterranean (Cartes & Carrassón, 2004). The increased abundance of gastropods, echinoderms, polychaetes – chiefly Eunicidae, sipunculans and chaetognaths in the stomachs, confirms that this species in the Greek Ionian Sea could be considered a "slow hunter", foraging mainly on organisms that live completely or partially buried in the substratum. The macrophyte consumption was rare in both sexes and probably reflects availability in the marine environment.

The data of the present study confirm that *A. antennatus* could be considered a less active and slower hunter than the other aristeid species (*A. foliacea*) found in the same area (Kapiris et al., 2010) and preys on detrivores or small predators occupying a lower position in the benthopelagic food chain (Maynou & Cartes, 1997). The feeding activity patterns of *A. antennatus* in the Greek Ionian Sea are, more or less, comparable to those reported in other geographical regions, such as the central (e.g. Relini & Orsi Relini, 1987; Follesa et al., 2009) or in the western Mediterranean (Cartes & Sardà, 1989; Maynou & Cartes, 1998). Apparent differences in the activity patterns should be attributed to the more oligotrophic character of the Ionian Sea (E. Mediterranean) in relation to the western one and to the bottom morphology (Cartes, 1995). The above mentioned oligotrophic character of the eastern Mediterranean could explain the presence of the increased number of some pelagic preys in its stomachs, in comparison to the western one (Cartes, 1995), but - as we said before - these preys constitute the minority comparing to the benthic ones. Some remains of the sympatric *A. foliacea* in the stomachs of *A. antennatus* and vice versa could be accidental, since they have been found in the sampling stations where both species coexisted and, thus, some body appendages were destroyed and mixed during the net tow (net feeding). It is possible that the smaller individuals of each species, due to their voracious character, can be fed by the adults of the other one. In any case, further study is necessary.

Only a partial differentiation in the feeding behaviour between sexes, in terms of both diet composition and feeding activity, is observed. Males exhibit lower values of fullness, food quality indices and evenness than females. Both sexes consume the same prey items, but in different abundance and occurrence. From the above results, a slightly higher predatory ability of females is shown. These differences could also be attributed to sexual dimorphism and to size difference between the sexes.

4.3.2 Seasonal differences

Taking into consideration the narrow depth sampling range, the estimated values of the trophic overlap indicate that, Season could not be considered as the main factor affecting the diet of blue-red shrimp in the Greek Ionian Sea, like in *A. foliacea*. Notwithstanding, some particular topics are analyzed below. The existence of regular Seasonal rhythms in the feeding activity of deep-water species is mainly related to the Seasonal fluctuations of abundance of prey they consume, the depth, the local geographical characteristics, the submarine canyons, the type of bottom, the Seabed, the Seasonal horizontal and diurnal vertical migrations, etc. (Cartes, 1993a, 1998). In addition to this, the Seasonal changes in stomach fullness of blue-red shrimp could be possibly linked to the oceanographic processes and to the several water masses, at least in the W. Mediterranean (Cartes et al., 2008b; Maynou, 2008).

The above slight Seasonal changes in the feeding dynamics of this aristeid in the Greek Ionian Sea seem to be related mainly to their biological processes (e.g. mating and reproduction) and to the food availability. The increased values of food quality indices and diversity support the finding that feeding activity seemed to increase qualitatively - in the period spring–summer, mainly for females. In addition to this, the observed highest empty stomachs found in these Seasons, mainly for females, could be attributed to the increased volume of the gonads which press the stomach. This increase of the highly energetic diet could be attributed to the increased pre- and reproductive activity observed in this period (Kapiris and Thessalou-Legaki, 2006, 2009). As Cartes et al. (2008a) noted *A. antennatus* seemed to increase the energy intake in its diet from February to April-June in the western Mediterranean. During winter both sexes of *A. antennatus* in Greek Ionian Sea consume an increased number of prey items, having as a result the highest stomach fullness, but of decreased quality. This phenomenon could be related to the mating period which takes place in this Season (Kapiris & Thessalou-Legaki, 2006, 2009). Besides the Seasonal feeding adaptation to the biological requirements, the food availability also plays an important role for this species in the Greek Ionian Sea. Madurell & Cartes (2005) point out that, in the same study area, the suprabenthos fauna (mysids, cumaceans, amphipods, isopods, and tanaidaceans) showed the highest densities in spring, while the zooplankton fauna (chiefly copepods, ostracods and chaetognaths) was more abundant in autumn and summer. In agreement with the results of the present study, the above fluctuations of food availability are also shown in the diet of *A. antennatus*. Thus, the diet of the blue-red shrimp probably reflects localized forage assemblages rather than a preference for specific items. In addition to this, these results reinforce the opinion concerning the "accidental hunting" of *A. antennatus*.

4.3.3 Ontogenetic differences

Comparison of the diet composition, dietary diversity and feeding activity between the small size, medium size and - only in females - large size individuals reveals that this species undergoes changes in feeding habits with increasing body size and gonad maturity in the Greek Ionian Sea. Small immature individuals consume less prey, mainly epibenthic, - but of increased quality - due to their smaller stomach. Larger mature specimens of both sexes are more efficient predators because of their greater swimming ability and their larger mandibles. The positive trend between increasing females' body size and consumption of larger prey is observed could be attributed to the population structure and to the morphological characteristics of the different size classes and sexes. In general, somatic growth and gonad development induce a change of *A. antennatus* feeding behaviour in the Greek Ionian Sea: as the body grows, an increasing mean weight of prey and mean number of prey items per stomach was obvious. However, almost the same prey occurred in the stomachs of small, medium and large specimens, but in different proportions.

5. Conclusions

Our results on the feeding ecology of both deep water shrimps could be considered as primary importance for the future management of deep water assemblages, since they play an important role. Since the deep waters in the E. Ionian Sea are almost unexploited, the present data could elucidate the relationships between species in this ecosystem improving, thus, the knowledge and the trophic relationships among the species helping in their integrated management in the future.

According all the studies carried out on both decapods feeding habits, *A. foliacea* exploits different resources from those used by *A. antennatus* and, despite both shrimps have similar morphologies and size ranges, the exploitation of different resources probably both species to coexist in the same areas (Cartes, 1995). In addition to this, since both deep-sea red shrimps belonging in the same family, have an almost similar depth distribution It is expected that they have similar energy values (in terms of wet mass), water body content (K. Kapiris unpublished observations) and oxygen consumption rates (Company & Sardà 1998).

Concluding, the increased demand of the large energetic content and the food availability in the same period make us suggest that both facts could stimulate fecundity in the deep-sea blue–red shrimp in the E. Mediterranean. A similar trend has been shown for the same species in the western Mediterranean (Cartes et al., 2008a, b). Generally, energy reserves strongly affect fecundity and reproduction in fishes (e.g. Lloret et al., 2005) and have been also observed in deep-water decapods (Fanelli & Cartes, 2008).

6. References

Aquastudio (1996). *Survey of red shrimp fishing in the Western Italian basins.* Final Report. CE DG XIV, Contract n° MED92/005.

Atkinson, D.B. (1995). The biology and fishery of roundnose grenadier (*Coryphaenoides rupestris* Gunnerus, 1765) in the north west Atlantic. In *Deep-water fisheries of the south Atlantic Oceanic Slope,* Hopper, A. G. (Ed.), NATO Asi. Series E., Applied Sciences, Vol. 296, pp. 51-111, Kluwer Academic Publishers, Dorbrecht.

Bianchini, M.L. & Ragonese, S. (1994). Life cycles and fisheries of the deepwater red shrimps *A. foliacea* and *A. antennatus. Proceedings of the International workshop held in the Istituto di Tecnologia della Pesca e del Pescato,* pp. 1-87, Mazara del Vallo. N.T.R.-I.T.P.P. Special Publication.

Bianchi, N. & Morri, C. (2000). Marine biodiversity of the Mediterranean Sea, situation, problems and prospects for future research. *Mar. Poll. Bull.* Vol. 40, No 5, pp. 367-376.

Bouchet, P.H. & Taviani, M. (1992). The Mediterranean deep sea fauna, pseudopopulations of Atlantic species?. *Deep-Sea Res.,* Vol. 39, No 2, pp. 169-184.

Brian, A. (1931). La biologia del fondo a "scampi" nel Mare Ligure. V. *Aristaeomorpha, Aristeus* ed altri macruri natante. *Boll. Mus. Zool. Anat. Comp. R. Univ. Genova,* Vol. 2, No 45, pp. 1-6.

Burukovsky, R.N., Romensky, L.L., Kozyaistva, R. & Okeanografii, I. (AtantNIRO) (1972). On the variability of the rostrum in the *Aristeus varidens* (Decapoda, Penaeidae). *Trudy Atlanticheskii Nauchna-issledovatel'skii Inst.* Vol. 42, pp. 156-161 [In Russian].

Carrassón, M. & Cartes, J.E. (2002). Trophic relationships in a Mediterranean deep sea fish community: partition of food resources, dietary overlap and comments within the Benthic Boundary Layer. *Mar. Ecol. Prog. Ser.*, Vol. 241, pp. 41-55.

Cartes, J.E. & Sardà, F. (1989). Feeding ecology of the deep-water aristeid crustacean *A. antennatus. Mar. Ecol. Prog. Ser.*, Vol. 54, pp. 229-238.

Cartes, J.E. & Sardà, F. (1992). Abundance and diversity of decapod crustaceans in the deep-Catalan Sea (Western Mediterranean). J. Natural Hist., Vol. 26, pp. 1305–1323.

Cartes, J.E. (1993a). Day-night feeding by decapod crustaceans in a deep-water bottom community in the Western Mediterranean. *J. Mar. Biol. Assoc. UK*, Vol. 73, pp. 795-811.

Cartes, J.E. (1993b). Deep-sea decapods fauna of the western Mediterranean: Bathymetric distribution and biogeographic aspects. *Crustaceana*, Vol. 65, No. 1, pp. 29-40.

Cartes, J.E. (1994). Influence of depth and season on the diet of the deep-water aristeid *Aristeus antennatus* along the continental slope (400 to 2300 m) in the Catalan Sea (Western Mediterranean). *Mar. Biol.*, Vol. 120, pp.639-648.

Cartes, J.E. (1995). Diets of, trophic resources exploited by, bathyal Penaeoidean shrimps from the Western Mediterranean. *Mar. Freshwater Res.*, Vol. 46, pp. 889-896.

Cartes, J.E. (1998) Feeding strategies and partition of food resources in deep-water decapod crustaceans (400–2300 m). *J. Mar. Biol. Assoc. UK.*, Vol. 78, pp. 509–524.

Cartes, J.E. & Maynou, F. (1998). Food consumption by bathyal decapod crustacean assemblages in the western Mediterranean: predatory impact of megafauna and the food consumption-food supply balance in a deep-water food web. *Mar. Ecol. Prog. Ser.*, Vol. 171, pp. 233-246.

Cartes, J.E. & Carrassón, M. (2004) Influence of trophic variables on the depth range distributions and zonation rates of deep-sea megafauna: the case of the Western Mediterranean assemblages. *Deep-Sea Res I*, Vol. 51, pp.263–279.

Cartes, J.E., Madurell, T., Fanelli, E. & López-Jurado, J.L. (2008a). Dynamics of suprabenthos zooplankton communities around the Balearic Islands (NW Mediterranean): influence of environmental variables and effects on higher trophic levels. *J. Mar. Syst.*, Vol. 71, pp. 316–335.

Cartes, J.E., Papiol, V. & Guajardo, B. (2008b). The feeding and diet of the deep-sea shrimp *A. antennatus* off the Balearic Islands (Western Mediterranean): influence of environmental factors and relationshipwith the biological cycle. *Prog. Ocean*, Vol. 79, pp. 37–54.

Cartes, J.E., Fanelli, E., Papiol, V. & Maynou, F. (2010). Trophic relationships at intrannual spatial and temporal scales of macro and megafauna around a submarine canyon off the Catalonian coast (western Mediterranean). *J. Sea Res.*, Vol. 63, pp. 180–190.

Chartosia, N., Tzomos, T.H., Kitsos. M.S., Karani. I., Tselepides. A. & Koukouras A. (2005). Diet comparison of the bathyal shrimps, *Aristeus antennatus* (Risso, 1816) and *Aristaeomorpha folicea* (Risso, 1827) (Decapoda, Aristeidar) in the Eastern Mediterranean. *Crustaceana* Vol. 78, No. 3, pp. 273-284.

Company, J.B. (1995) Estudi comparatiu de les estratègies biològiques dels crustacis decàpodes de la Mar Catalana. Ph.D. thesis, University of Barcelona.

Company. J.B. & Sardà, F. (1998) Metabolic rates and energy content of deep-sea benthic decapod crustaceans in the Western Mediterranean Sea. *Deep-sea Res. Part I, Oceanographic Research Papers*, Vol. 45, pp. 1861–1880.

Company, J.B., Maiorano, P., Tselepides, A., Politou, C.-Y., Plaiti, W., Rotllant, M. & Sardà, F. (2004). Deep-sea decapod crustaceans in the western and central Mediterranean Sea: preliminary aspects of species distribution, biomass and population structure. *Sci. Mar.*, Vol. 68 (Suppl. 3), pp. 73-86.

Dall, W. (1968). Food and feeding of some Australian penaeids shrimps. *FAO Fisheries Report Series,* Vol. 57, No. 2, pp. 251-258.

Dall, W.B., Hill, J., Rothlisberg, P.C. & Sharples, D.J. (1990). *The biology of the Penaeidae.* London, Academic Press, pp 489.

Dallagnolo R., Perez J.A.A., Pezzuto P.R. & Wahrlich R. (2009) The deep-sea shrimp fishery off Brazil (Decapoda: Aristeidae) development and present status. *Latin American Journal of Aquatic Research,* Vol. 37, pp. 327–346.

Danovaro, R., Company, J.B., Corinaldesi, C., D'Onghia, G., Galil, B., Gambi1, C., Gooday, A.J., Lampadariou, N., Luna, G.M., Morigi, C., Olu, K., Polymenakou, P., Ramirez-Llodra, E., Sabbatini, A., Sardà F., Sibuet, M. & Tselepides, A. (2010). Deep-sea Biodiversity in the Mediterranean Sea: The Known, the Unknown, and the Unknowable. *PLoS ONE,* Vol. 5, No. 8, 25 p.

Demestre, M. & Martín, P. (1993). Optimum exploitation of a demersal resource in the Western Mediterranean, the fishery of the deep-water shrimp *A. antennatus* (Risso, 1816). *Sci. Mar.,* Vol. 57, No 2, pp. 175-182.

Demestre, M. (1994). Fishery and population dynamics of *A. antennatus* on the Catalan coast (NW Mediterranean). In *Life cycles and fisheries of the deep-water red shrimps A. foliacea and A. antennatus,* Bianchini & Ragonese (Eds), N.T.R.-I.T.P.P., Spec. Publ., Vol.3, pp. 19-20.

D'Onghia, G., Matarrese, A., Tursi, A. & Maiorano, P. (1994). Biology of *A. antennatus* and *Aristaeomorpha foliacea* in the Ionian Sea (Central Mediterranean). In *Life cycles and fisheries of the deep-water red shrimps A. foliacea and A. antennatus*, Bianchini, M. L., Ragonese, S. (Eds),, N.T.R.-I.T.P.P., Spec. Publ., 3, pp. 55-56.

D'Onghia, G., Maiorano, P., Matarrese, A. & Tursi, A. (1998). Distribution, biology, and population dynamics of *A. foliacea* (Risso, 1827) (Decapoda, Natantia, Aristeidae) in the North-Western Ionian Sea (Mediterranean Sea). *Crustaceana,* Vol. 71, No 5, pp. 18-544.

D'Onghia, G., Politou, C.Y., Bozzano, A., Lloris, D., Rotllant, G., Sion, L., Mastrototaro, F. (2004). Deep-water fish assemblages in three areas of the Mediterranean Sea. *Sci. Mar.,* Vol. 68, No. 3, pp. 87–99.

D'Onghia, G., Capezzuto, F., Mytilineou, Ch., Maiorano, P., Kapiris, K., Carlucci, R., Sion, L. & Tursi, A. (2005). Comparison of the population structure and dynamics of *A. antennatus* (Risso, 1816) between exploited and unexploited areas in the Mediterranean Sea. *Fish. Res.,* 2005 Vol. 76, pp. 23-38.

Emig, C.C. & Geistdoerfer, P. (2004). The Mediterranean deep-sea fauna, historical evolution, bathymetric variations and geographical changes. Carnets de Geologie/Notebooks on Geology, Maintenon, Article 2004/01 (CG2004_A01_CCE-PG).

Fanelli, E. & Cartes, J.E. (2008). Spatio-temporal changes in gut contents and stable isotopes in two deep Mediterranean pandalids: influence on the reproductive cycle. *Mar Ecol Prog Ser.,* Vol. 355, pp. 219-233.

Follesa, M. C., Cuccu, D., Murenu, M., Sabatini, A. & Cau, A. (1998). Aspetti riproduttivi negli Aristeidi, *A. foliacea* (Risso, 1827) e *A. antennatus* (Risso, 1816), della classe di eta 0+ e 1+. *Biol. Mar. Medit.*, Vol. 5, No. 2, pp. 232-238.

Follesa, M.C., Porcu, C., Gastoni, A., Mulas, A., Sabatini, A. & Cau, A. (2009). Community structure of bathyal decapod crustaceans off South-Eastern Sardinian deep-waters (Central-Western Mediterranean). *Mar. Ecol.*, Vol. 30, pp. 188–199.

Fredj, G. & Laubier, L. (1985) The deep Mediterranean benthos. In: *Mediterranean marine ecosystems,* Moraitou-Apostolopoulou & Kiortsis (Eds), pp. 109-146, Plenum Press, New York

Gartner, J.V.Jr., Crabtree R.E. & Sulak, K.J. (1997). Feeding at depth. In: *Deep-sea fishes,* Randall & Farrell (Eds.), pp. 115-193, Academic Press, San Diego.

Ghidalia, W. & Bourgois, F. (1961). Influence of temperature and light on the distribution of shrimps in medium and great depths. *Stud. Gen. Fish. Coun. Medit.*, Vol. 16, pp. 1-49.

Gönülal, O., Özcan, T. & Katagan, T. (2010). A contribution on the distribution of the giant red shrimp *A. foliacea* (Risso, 1827) along the Aegean sea and Mediterranean part of Turkey. *Rapp. Comm. int. Mer Médit.*, Vol. 39, p. 534.

Gracia, A., Vázquez-Bader, A.R., Lozano-Alvarez, E. & Briones-Fourzán, P. (2010) Deep-water shrimp (Crustacea:Penaeoidea) off the Yucatan peninsula (southern Gulf of Mexico): a potential fishing resource. *J. Shellfish Res.* Vol. 29, pp. 37–43.

Gristina, M.F., Badalamenti, F., Barbera, G., D'Anna, G. & Pipitone C. (1992). Preliminary data on the feeding habits of *A. foliacea* (Risso) in the Sicilian Channel. *Oebalia* suppl. XVII, pp. 143-144.

Héroux, D. & Magnan, P. (1996) In situ determination of food daily ration in fish: review and field evaluation. *Environ. Biol. Fish.*, Vol. 46, 61–74.

Hiller-Adams, P. & Childress, J.J. (1983). Effects of feeding history and food deprivation on respiration and excretion rates of the bathypelagic mysid *Gnathophausia ingens. Biol Bull,* Vol. 165, pp. 182-196.

Hopkins, T.S. (1985). Physics of the sea. In: *Key Environments: Western Mediterranean,* Margalef R. (Ed.), pp. 100-125, Pergamon Press, New York.

Hyslop, E.J. (1980) Stomach contents analysis - a review of methods and their application. *J. Fish Biol,* Vol. 17, pp. 411–429.

Jumars, P.A. & Gallagher, E.D. (1982). Deep-sea community structure: three plays on the benthic proscenium. In: *The environment of the deep sea,* Ernst, W.G. & J.G. Morin (Eds.), Prentice Hall, Englewood Clfts.

Kallianiotis, A., Sophronidis, K., Vidoris, P., Tselepides, A. (2000). Demersal fish and megafaunal assemblages on the Cretan continental shelf and slope (NE Mediterranean), seasonal variation in species density, biomass and diversity. *Prog. Oceanogr.*, Vol. 46, pp. 429-455.

Kapiris, K., Thessalou-Legaki, M., Moraitou-Apostolopoulou, M., Petrakis, G. & Papaconstantinou, C. (1999). Population characteristics and feeding parameters of *A. foliacea* and *Aristeus antennatus* (Decapoda: Aristeidae) from the Ionian Sea (Eastern Mediterranean). In: *The Biodiversity crisis and crustacea. Crustacean Issues,* Vol. 12, pp. 177-191.

Kapiris, K. & Thessalou-Legaki, M. (2001a). Sex-related variability of rostrum morphometry in *A. antennatus* from the Ionian Sea (Eastern Mediterranean). *Hydrobiologia,* Vol. 449, pp. 123-130.

Kapiris, K. & Thessalou-Legaki, M. (2001b). Observations on the reproduction of *A. foliacea* (Crustacea, Aristeidae) in the SE Ionian Sea. *Rapp. Comm. Int. Mer. Médit.*, Vol. 36, p. 281.

Kapiris, K., Tursi, A., Mytilineou, Ch., Kavadas, S., D' Onghia, G. & Spedicato, M. T. (2001c). Geographical and bathymetrical distribution of *A. foliacea* and *A. antennatus* (Decapoda, Aristeidae) in the Greek Ionian Sea. *Rapp. Comm. Int. Mer. Médit.*, Vol. 36, p. 280.

Kapiris, K., Moraitou-Apostolopoulou, M. & Papaconstantinou, C. (2002). The growth of male secondary sexual characters in *A. foliacea* and *A. antennatus* (Decapoda, Aristeidae) in the Greek Ionian Sea (Eastern Mediterranean). *J Crustacean Biology*, Vol. 22, No. 4, pp. 784-789.

Kapiris, K. (2004). Biology and fishery of the deep water shrimps *A. foliacea* (Risso, 1827) and *A. antennatus* (Risso, 1816) (Decapoda: Dendrobranchiata). Ph. D. thesis, University of Athens, 289 pp.

Kapiris, K. (2005). Morphometric structure and allometry profiles of the giant red shrimp *A. foliacea* (Risso, 1827) in the eastern Mediterranean. *J. Nat. Hist.*, Vol. 39, No. 17, pp. 1347-1357.

Kapiris K., Thessalou-Legaki M. (2006). Comparative fecundity and oocyte size of *Aristaeomorpha foliacea* and *Aristeus antennatus* in the Greek Ionian Sea (E. Mediterranean) (Decapoda: Aristeidae). *Acta Zoologica*, Vol. 87, pp. 239-245.

Kapiris, K., Kavvadas, S. (2009). Morphometric structure and allometry profiles of the red shrimp *A. antennatus* (Risso, 1816) in the Eastern Mediterranean. *Aquat. Ecol.*, Vol. 43, No. 4, pp. 1061-1071.

Kapiris, K. & Thessalou-Legaki, M. (2009). Comparative Reproduction Aspects of the Deep-water Shrimps *A. foliacea* and *A. antennatus* (Decapoda, Aristeidae) in the Greek Ionian Sea (Eastern Mediterranean). *Int. J. Zool.*, Vol. 2009, Article ID 979512, 9 pages, doi,10.1155/2009/979512.

Kapiris, K., Thessalou-Legaki, M., Petrakis, G. & Conides, A. (2010). Ontogenetic shifts and temporal changes the trophic patterns of deep-sea red shrimp *A. foliacea* (Decapods, Aristeidae) in the E. Ionian Sea (E. Mediterranean). *Mar. Ecol.*, Vol. 31, No. 2, 341-354.

Kapiris, K. & Thessalou-Legaki, M. (2011). *Feeding ecology of the deep-sea red shrimp Aristeus antennatus in the NE Ionian Sea (E. Mediterranean).* J. Sea Res., Vol. 65, pp. 151-160.

Klein, B., Roether, W., Manca, B.B., Bregant, D., Beitzel, V., Kovacevic, V. & Luchetta, A. (1999). The large deep water transient in the Eastern Mediterranean. *Deep-sea Res., Vol. 46, pp. 371-414.*

Kuttyamma, V.J. (1974). Observations on the food and feeding of some penaeid prawns of Cochin area. Journal *J. Mar. Biol. Assoc. of India,* Vol. 15, pp. 189–194.

Labropoulou, M. & Papaconstantinou, C. (2000). Community structure of deep-sea demersal fish in the North Aegean Sea (northeastern Mediterranean). *Hydrobiology*, Vol. 440, pp. 281–296.

Lagardère, J.P. (1977). Recherches sur la distribution verticale et sur l'alimentation des crustacés décapodes benthiques de la pente continentale du Golfe de Gascogne. *Bull. Cnt. Etud. Rech. Scient. Biarritz*, Vol. 11, No. 4, pp. 367-440.

Levi, D. & Vaccchi, M.J. (1988). Macroscopic scale for simple and rapid determination of sexual maturity in *A. foliacea* (Risso, 1826) (Decapoda, Penaeidae). *Crustacean Biol,*.Vol. 8, No. 4, pp. 532-538.

Lloret, J., Galzin, R., Gil de Sola, L., Souplet, A. & Demestre, M. (2005). Habitat related differences in lipid reserves of some exploited fish species in the north-western Mediterranean continental shelf. *J Fish Biol.*, Vol. 67, pp. 51-65.

Macpherson, E. (1980). Regime alimentaire de *Galeus melastomus* Rafinesque, 1810, *Etmopterus spinax* (L., 1758) et *Scymnorhinchus licha* (Bonnaterre, 1788) en Mediterranee occidentale. *Vie Milieu*, Vol. 30, pp. 139-148.

Madurell, T. (2003). Feeding strategies and trophodynamic requirements of deep-sea demersal fish in the Eastern Mediterranean. Ph. D. Thesis, Universitat de les Illes Balears, pp. 1-251.

Madurell, T., Cartes, J.E. & Labropoulou, M. (2004). Changes in the structure of fish assemblages in a bathyal site of the Ionian Sea (eastern Mediterranean). *Fish. Res.*, Vol. 66, pp. 245–260

Madurell, T. & Cartes, J.E. (2005) Temporal changes in feeding habits and daily rations of *Hoplostethus mediterraneus*, Cuvier, 1829 in the bathyal Ionian Sea (eastern Mediterranean). *Mar Biol.*, Vol. 146, pp. 951–962.

Madurell, T. & Cartes, J.E. (2006). Trophic relationships and food consumption of slope dwelling macrourids from the bathyal Ionian Sea (eastern Mediterranean). *Mar. Biol.*, Vol. 148, pp. 1325–1338.

Matarrese, A., D'Onghia, G., De Florio, M., Panza, M. & Constantino, G. (1995). Recenti acquisizioni sulla distribuzione batimetrica di *A. foliacea* ed *A. antennatus* (Crustacea, Decapoda) nel Mar Ionio. *Biol. Mar. Medit.*, Vol. 2, No. 2, pp. 299-300.

Matarrese, A., D'Onghia, G., Tursi, A. & Maiorano, P. (1997). Vulnerabilità e resilienza di *Aristeomorpha foliacea* (Risso, 1816) and *A. antennatus* (Risso, 1816) (Crostacei, Decapodi) nel Mar Ionio. *SITE Atti*, Vol. 18, pp. 535-538.

Mauchline, J. & Gordon, J.D.M. (1984). Diets and bathymetric distributions of the macrurid fish of the Rockall Trough, northeastern *Atlantic Ocean. Mar. Biol.*, Vol. 81, pp. 107-121.

Mauchline, J. & Gordon, J.D.M. (1986). Foraging strategies of deep-sea fish. *Fish Biol.*, Vol. 26, pp. 527-535.

Maurin, C. & Carries, C. (1968). Note préliminaire sur l'alimentation des Crevettes profondes. *Rapp. Comm. Int. Mer Médit.*, Vol. 19, No. 2, pp. 155-156.

Maynou, F. & Cartes, J. (1997). Field estimation of daily ratio in deep-sea shrimp *Aristeus antennatus* (Crustacea: Decapoda) in the Western Mediterranean. *Mar. Ecol. Prog. Ser.*, Vol. 153, pp. 191-196.

Maynou, F. & Cartes, J. (1998). Daily ration estimates and comparative study of food consumption in nine species of deep-water decapod crustaceans of the NW Mediterranean. *Mar. Ecol. Prog. Ser.*, Vol. 171, pp. 221-231.

Maynou, F. & Cartes, J.E. (2000). Community structure of bathyal decapod crustacean assemblages off the Balearic Islands (south-western Mediterranean). *J. Mar. Biol. Assoc. UK*, Vol. 80, pp. 789-798.

Maynou, F. (2008). Environmental causes of the fluctuations of blue-red shrimp (*Aristeus antennatus*) landings in the Catalan Sea. *J Marine Syst.*, Vol. 71, pp. 294-302.

Monniot, C. & Monniot, F. (1990). Revision of the class Sorberacea (benthic tunicates) with descriptions of seven new species. *Zool. J. Linn. Soc.*, Vol. 99, pp. 239-290.

Moriarty, D.J.W. & Barclay, M.C. (1981). Carbon and nitrogen content of food and the assimilation efficiencies of penaeid prawns in the Gulf of Carpentaria. *Aust J Mar Fresh Res*, Vol. 32, pp. 245-251.

Mura, M., Orru, F. & Cau, A. (1997). Osservazioni sull' accrescimento di individui fase pre-riproduttiva di *A. antennatus* e *A. foliacea*. *Biol. Mar. Medit.*, Vol. 4, No. 1, pp. 254-261.

Mura, M., Saba, S. & Cau, A. (1998). Feeding habits of demersal aristeid. *Anim. Biol.*, Vol. 7, pp. 3-10.

Mytilineou, Ch., Kavadas S., Politou C.-Y., Kapiris K., Tursi A. & Maiorano P. (2006). Catch composition in red shrimp (*A. foliacea* and *A. antennatus*) grounds in the eastern Ionian Sea. *Hydrobiologia*, Vol. 557, pp. 155-160.

Nouar, A. (2001) *Rapp.* Bio-écologie d'A. *antennatus* (Risso, 1816) et de *Parapenaeus longirostris* (Lucas, 1846) des côtes algériennes. *Comm. Int. Mer. Médit.*, Vol. 36, p. 304.

Orsi Relini, L. & Semeria, M. (1983). Oogenesis and fecundity in bathyal penaeid prawns, *A. antennatus* and *A. foliacea*. *Rapp. Verb. Reun.*, CIESM, Vol. 28, No. 3, pp. 281-284.

Orsi Relini, L. & Relini, G. (1998). Seventeen instars of adult life in female *A. antennatus* (Crustacea, Decapoda, Aristeidae). A new interpretation of life span and growth. *J. Nat. Hist.*, Vol. 32, pp. 719-1734.

Papaconstantinou, C. & Kapiris, K. (2001). Distribution and population structure of the red shrimp (*Aristeus antennatus*) on an unexploited fishing ground in the Greek Ionian Sea. Aquat. Living Resources, Vol. 14, No. 5, pp. 303-312.

Papaconstantinou, C. & Kapiris, K. (2003) The biology of the giant red shrimp (*A. foliacea*) at an unexploited fishing ground in the Greek Ionian Sea. *Fish. Res.*, Vol. 62, pp. 37-51.

Paramo, J. & Ulrich, S.P. (2011). Deep-sea shrimps *A. foliacea* and *Pleoticus robustus* (Crustacea: Penaeoidea) in the Colombian Caribbean Sea as a new potential fishing resource. *J. Mar. Biol. Assoc. UK*, pp. 1-8, doi:10.1017/S0025315411001202.

Pérez Farfante, I. & Kensley, B. (1997). Penaeoid and sergestoid shrimps and prawns of the world. Keys and diagnoses for the families and genera. *Mém. Mus. Nat. Hist. Nat.*, Vol. 175, pp. 1-233.

Politou, C.-Y., Kavadas, S., Mytilineou, Ch., Tursi, A., Lembo, G. & Carlucci, R.J. (2003). Fisheries resources in the deep waters of the Eastern Mediterranean (Greek Ionian Sea). *Northw. Atl. Fish. Sci.*, Vol. 31, pp. 35-46.

Politou, C.-Y., Kapiris, K., Maiorano, P., Capezzuto, F. & Dokos, J. (2004). Deep-Sea Mediterranean biology, the case of *A. foliacea* (Risso, 1827) (Crustacea, Decapoda, Aristeidae). *Sci. Mar.*, Vol. 68 (Suppl. 3), pp. 117-127.

Politou, C-Y., Maiorano, P., D'Onghia, G. & Mytilineou, Ch. (2005). Deep-water decapod crustacean fauna of the Eastern Ionian Sea. *Belg. J. Zool.* , Vol. 135, No. 2, pp. 235-241.

Polunin, N.V.C., Morales-Nin, B., Herod, W., Cartes, J.E., Pinnegar, J.K. & Moranta, J. (2001). Feeding relationships in Mediterranean bathyal assemblages elucidated by carbon and nitrogen stable-isotope data. *Mar. Ecol. Prog. Ser.*, Vol. 220, pp. 13–23.

Puig, P., Company, J.B., Sardà, F. & Palanques, A. (2001). Responses of deep-water shrimp populations to the presence of intermediate nepheloid layers on continental margins. *Deep-Sea Res.*, Vol. 48, pp. 2195-2207.

Ragonese, S. (1995). Geographical distribution of *A. foliacea* (Crustacea, Aristeidae) in the Sicilian Channel (Mediterranean Sea). *ICES Mar. Sci. Symp.*, pp. 183-188.

Ragonese, S., Bianchini, M.L. & Gallucci, V.F. (1994a). Growth and mortality of the red shrimp *A. foliacea* in the Sicilian Channel (Mediterranean Sea). *Crustaceana*, Vol. 67, pp. 348-361.

Ragonese, S., Bianchini, M.L., Di Stefano, L., Campagnuolo, S. & Bertolino, F. (1994b). *A. antennatus* in the Sicilian Channel In *Life cycles and fisheries of the deep-water red shrimps A. foliacea and A. antennatus*, Bianchini & Ragonese, (Eds.), N.T.R.-I.T.P.P., Spec. Publ., Vol. 3, p. 44.

Ramirez-Llodra, E., Company, J. B., Sardà F. & Rotllant, G. (2009). Megabenthic diversity patterns and community structure of the Blanes submarine canyon and adjacent slope in the Northwestern Mediterranean, A human overprint? *Mar. Ecol.*, pp. 1-16.

Randall, D.J. & Farrell, A.P. (Eds.). (1997). *Deep-sea fishes*. Academic Press, N.Y.

Relini Orsi, L. & Wurtz, M. (1977). Aspetti della rete trofica batiale riguardanti *Aristeus antennatus*. *Atti. IX Congr. Naz. Soc. Ital. Biol. Mar.*, pp. 389-398.

Relini, G., Orsi Relini, L. (1987). The decline of red shrimps stocks in the Gulf of Genoa. *Inv. Pesq.* 1987, Vol. 51(Suppl. 1), pp. 245-260.

Relini, M., Maiorano, P., D'Onghia, G., Orsi Relini, L., Tursi, A. & Panza, M. (2000). A pilot experiment of tagging the deep shrimp *A. antennatus* (Risso, 1816). *Sci. Mar.*, Vol. 64, No. 3, pp. 357-361.

Ruffo, S. (1998). The Amphipoda of the Mediterranean. Part 4. *Mém. Inst. Océanog. Monaco*, Vol. 13, pp. 1-959.

Sardà, F. & Cartes, J.E. (1993). Distribution abundance and selected biological aspects of *A. antennatus* in deep-water habitats in NW Mediterranean. *B.I.O.S.*, Vol. 1, No 1, pp. 59-73.

Sardà, F., Cartes, J.E. & Company, J.B. (1994a). Spatio-temporal variations in megabenthos abundance in three different habitats of the Catalan deep-sea (Western Mediterranean). *Mar. Biol.*, Vol. 120, pp. 211-219.

Sardà, F. & Cartes, J.E. (1994b). Status of the qualitative aspects in *A. antennatus* fisheries in the North Western Mediterranean. In: *Life cycles and fisheries of the deep-water red shrimps A. foliacea and A. antennatus*, Bianchini, M.L., Ragonese, S. (Eds.), N.T.R.-I.T.P.P., Spec. Publ., 3, p. 23-24.

Sardà, F. & Cartes, J.E. (1997). Morphological features and ecological aspects of early juvenile specimens of the aristeid shrimp *A. antennatus* (Risso, 1816). *Mar. Freshw. Res.*, Vol. 48, pp. 73-77.

Sardà, F., Maynou, F. & Talló, L. (1998). Seasonal and spatial mobility patterns of rose shrimp (*A. antennatus* Risso, 1816) in the western Mediterranean, results of a long-term study. *Mar. Ecol. Prog. Ser.* Vol. 159, pp. 133-141.

Sardà, F., Company, J.B. & Maynou, F. (2003). Deep-sea shrimp *A. antennatus* Risso, 1816 in the Catalan sea, a review and perspectives. *J. Northw. Atl. Fish. Sci.*, Vol. 31, pp. 127-136.

Sardà, F., Calafat, A., Mar Flexas, M, Tselepides, A., Canals, M., Espino, M. & Tursi, A. (2004). An introduction to Mediterranean deep-sea biology, *Sci. Mar.*, Vol. 68, pp. 7-38.

Stefanescu, C., Morales-Nin, B. & Massutí, E. (1994). Fish assemblages on the slope in the Catalan Sea (western Mediterranean): influence of a submarine canyon. *J. Mar. Biol. Assoc. UK.*, Vol. 74, pp. 499–512.

Tselepides, A. & Eleftheriou, A. (1992). South Aegean (eastern Mediterranean) continental slope benthos: macroinfaunal-environmental relationships. In: *Deep-sea Food Chains and the Global Carbon Cycle*, Rowe, G.T. & Pariente, V. (Eds.), pp. 139–156, Kluwer Academic Publisher, Dordrecht.

Tyler, P.A. & Zibrowius, H. (1992). Submersible observations of the invertebrate fauna on the continental slope southwest of Ireland. *Oceanol. Acta*, Vol. 15, pp. 211-226.

Williams, A.B. (1955). A contribution to the life histories of commercial shrimps (Penaeidae) in North Carolina. *Bull. Mar. Sci. of the Gulf and Caribbean*, Vol. 5, pp. 116-146.

Permissions

The contributors of this book come from diverse backgrounds, making this book a truly international effort. This book will bring forth new frontiers with its revolutionizing research information and detailed analysis of the nascent developments around the world.

We would like to thank Dr. Kapiris Kostas, for lending his expertise to make the book truly unique. He has played a crucial role in the development of this book. Without his invaluable contribution this book wouldn't have been possible. He has made vital efforts to compile up to date information on the varied aspects of this subject to make this book a valuable addition to the collection of many professionals and students.

This book was conceptualized with the vision of imparting up-to-date information and advanced data in this field. To ensure the same, a matchless editorial board was set up. Every individual on the board went through rigorous rounds of assessment to prove their worth. After which they invested a large part of their time researching and compiling the most relevant data for our readers. Conferences and sessions were held from time to time between the editorial board and the contributing authors to present the data in the most comprehensible form. The editorial team has worked tirelessly to provide valuable and valid information to help people across the globe.

Every chapter published in this book has been scrutinized by our experts. Their significance has been extensively debated. The topics covered herein carry significant findings which will fuel the growth of the discipline. They may even be implemented as practical applications or may be referred to as a beginning point for another development. Chapters in this book were first published by InTech; hereby published with permission under the Creative Commons Attribution License or equivalent.

The editorial board has been involved in producing this book since its inception. They have spent rigorous hours researching and exploring the diverse topics which have resulted in the successful publishing of this book. They have passed on their knowledge of decades through this book. To expedite this challenging task, the publisher supported the team at every step. A small team of assistant editors was also appointed to further simplify the editing procedure and attain best results for the readers.

Our editorial team has been hand-picked from every corner of the world. Their multi-ethnicity adds dynamic inputs to the discussions which result in innovative outcomes. These outcomes are then further discussed with the researchers and contributors who give their valuable feedback and opinion regarding the same. The feedback is then collaborated with the researches and they are edited in a comprehensive manner to aid the understanding of the subject.

Apart from the editorial board, the designing team has also invested a significant amount of their time in understanding the subject and creating the most relevant covers. They scrutinized every image to scout for the most suitable representation of the subject and create an appropriate cover for the book.

The publishing team has been involved in this book since its early stages. They were actively engaged in every process, be it collecting the data, connecting with the contributors or procuring relevant information. The team has been an ardent support to the editorial, designing and production team. Their endless efforts to recruit the best for this project, has resulted in the accomplishment of this book. They are a veteran in the field of academics and their pool of knowledge is as vast as their experience in printing. Their expertise and guidance has proved useful at every step. Their uncompromising quality standards have made this book an exceptional effort. Their encouragement from time to time has been an inspiration for everyone.

The publisher and the editorial board hope that this book will prove to be a valuable piece of knowledge for researchers, students, practitioners and scholars across the globe.

List of Contributors

Tomotaka Shiraishi
Wakayama Research Center of Agriculture, Forestry and Fisheries, Japan

Ryoma Kamikawa
University of Tsukuba, Japan

Yoshihiko Sako
Kyoto University, Japan

Ichiro Imai
Hokkaido University, Japan

Karola Böhme, Inmaculada C. Fernández-No, Jorge Barros-Velázquez and Pilar Calo-Mata
Department of Analytical Chemistry, Nutrition and Food Science, School of Veterinary Sciences, University of Santiago de Compostela, Lugo, Spain

Jose M. Gallardo
Department of Food Technology, Institute for Marine Research (IIM-CSIC), Vigo, Spain

Benito Cañas
Department of Analytical Chemistry, University Complutense of Madrid, Madrid, Spain

Mara Schuler and Petra Bauer
Dept. Biosciences-Plant Biology, Saarland University, Saarbrücken, Germany

S. Hassing
Faculty of Engineering, Institute of Technology and Innovation, University of Southern Denmark, Denmark

K.D. Jernshøj
Faculty of Science, Department of Biochemistry and Molecular Biology, Celcom, University of Southern Denmark, Denmark

L.S. Christensen
Kaleido Technology, Denmark

O. Al-Dayel, O. Al-Horayess, J. Hefni, A. Al-Durahim and T. Alajyan
King Abdulaziz City for Science and Technology, Riyadh, Saudi Arabia

Yuichi Sakamoto, Keiko Nakade, Naotake Konno and Toshitsugu Sato
Iwate Biotechnology Research Center, Japan

Keiko Nakade
TSUMURA & CO, Japan

Toshitsugu Sato
Kitami Institute of Technology, Japan

Kostas Kapiris
Hellenic Centre for Marine Research, Institute of Marine Biological Resources, Greece

Printed in the USA
CPSIA information can be obtained
at www.ICGtesting.com
JSHW011332221024
72173JS00003B/123

9 781632 393418